Criteria for Moisture Control

Criteria for Moisture Control

G.W. Brundrett, BEng, PhD, CEng, MIMechE, MCIBSE, FRSH, MASHRAE
Research Manager: Buildings and Environment,
The Electricity Council Research Centre

Butterworths
London   Boston   Singapore   Sydney   Toronto   Wellington

PART OF REED INTERNATIONAL P.L.C.

First published 1990
© Butterworth & Co. (Publishers) Ltd, 1990

British Library Cataloguing in Publication Data

Brundrett, G. W. (Geoffrey Wilmot)
  Criteria for moisture control.
  1. Buildings. Humidity
  I. Title
  697.92

ISBN 0-408-02374-0

Library of Congress Cataloging-in-Publication Data

Brundrett, G. W. (Geoffrey Wilmot)
  Criteria for moisture control/G. W. Brundrett.
    p.   cm.
  Includes bibliographical references.
  ISBN 0-408-02374-0
  1. Dampness in buildings.   2. Moisture.   I. Title.
TH9031.B78   1990
693.8'93—dc20   89-48285
                                                          CIP

Composition by Genesis Typesetting, Borough Green, Kent
Printed and bound in Great Britain by Butler & Tanner, Frome, Somerset

Preface

Count Rumford, the practical scientist, held two strong beliefs. The first was that science was only useful when applied. The second was that progress could only be measured if from time to time we record the state of the art. My purpose in writing this book is for the same two reasons. Moisture is essential to life and has many and varied influences on us. I have set out to show the complexity and subtlety of these influences which I personally find fascinating. The book is also meant to be the companion to the 'Handbook of Dehumidification' published in 1987. The first book outlined how you could control moisture. This one tells you why.

The layout of the chapters puts general principles at the beginning and detailed applications towards the end of the book. The chapters are designed to be self-contained for those who only seek an answer to one particular problem. The material is not new but by assembling it in this way I hope that the reader will find an interesting overview.

One of the delights in my professional life is the involvement in many aspects of the environment. Most of the specialists in this area are friendly and helpful in sharing their knowledge. I gratefully acknowledge the help I have received over the years from my colleagues. Whenever possible their references are identified in the text. However, special mention must be made of Dr Agnes Onions of the Commonwealth Mycological Institute who introduced me to moulds; Dr Maureen Blackmore and Dr Elaine Prisk of the Liverpool Polytechnic who worked on food and clothing with us; Mr Frank Jones and Mr Alan Godfree of Altwell for guiding me on microbiological appraisals; Dr Bernard Bailey of the National Institute of Agricultural Engineering and Mr John Weir of the Farm Electric Centre for help in plant growth; Dr David Wyon of the Swedish Building Research Institute for discussions on eye dryness; Dr Brian Moore of our own research centre for data on corrosion; and Mr Bill Baker, also of our own laboratory, for guidance on electrical matters.

Mrs Ruth Galvin, our librarian, searched diligently for the references. Mr Geoff Ratcliff, our Assistant Director, very helpfully suggested ways of improving the manuscript. This took much of his spare time and I do appreciate that sacrifice. Mrs Joan Hughes deserves special praise not just for her outstanding secretarial skills but for the gentle enthusiasm and willingness with which she turned script into polished pages.

Finally I must acknowledge the support from my family. The patient encouragement of my wife Janet, despite paper mountains of references growing all over our house, and the cheerfulness of our children, Timothy and Jane, made it all possible.

My thanks to all and my apologies in advance to those who have helped but whose names I have accidentally not included.

G.W. Brundrett

Contents

Chapter 1

Introduction

Water has always been recognized as a vital component of life. In the Western world the Greek philosopher Thales of Miletus around 580 BC led a campaign to abandon the mythological gods ruling at that time and replace them by natural forces. He believed that all substances were forms of water and that a natural cycle took earth, air and water through the bodies of plants and animals back to their forms of earth, air and water. This view persisted in Britain at least until the time of Newton in the seventeenth century.

In the Eastern world the Chinese believed in five elements, metal, wood, water, fire and earth, and understood that life comprised two opposing forces constantly changing their balance. These two forces were Yin and Yang. Yin represented passive dark forces and was associated with earth and water, while Yang represented the active bright penetrating forces and was associated with fire. This concept spread to Japan in the third century BC and strong traces remain today, particularly in selecting important days and unlucky days.

Now that science has led us to a much clearer concept of atomic physics we still find that water is extraordinary and anomalous in nearly all its properties. It is the least understood and most complex of all familiar substances. The atomic view of matter defines three components of an atom. At the centre of the atom resides the proton with a positive charge. Around this proton orbit electrons, which are very light in mass in comparison with the proton. Neutrons, similar to protons but without their charge, can also reside with the protons at the nucleus. The electrons stay within their orbits and there are several different orbits. The inner shell has a maximum capacity for two electrons. The next layer has a maximum capacity for eight. Water is a molecule comprising two hydrogen atoms and one oxygen atom. The hydrogen atom has one proton and one electron. Oxygen has eight protons, eight neutrons and eight electrons. Two of its electrons occupy the first orbit, the remaining six the second. Two electrons are needed to complete the second electron shell. Thus two hydrogen atoms, each with one electron, can combine with one oxygen atom which needs two electrons to complete its outer shell. The hydrogen atom in turn shares one electron from the oxygen atom. The water molecule can therefore be represented by *Figure 1.1.*

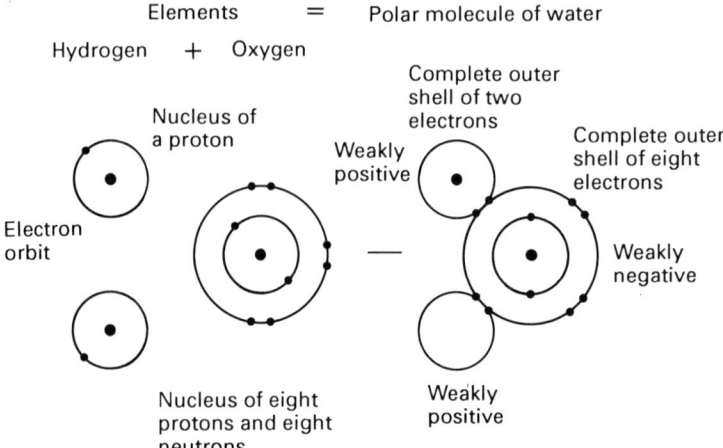

Figure 1.1 The electrons are shared (covalent) to complete the outer electron shell for both the hydrogen and the oxygen atoms when they form the water molecules

The result of hydrogen and oxygen combining is that the outer electrons are shared. The water molecule created by this is therefore covalent. It is not a linear molecule but is bent. When different elements combine in this way one atom will have a greater attraction for the electrons than the other. The oxygen atom has this greater electronegativity and therefore has a slight negative charge. This leaves the hydrogen atom with a slight positive charge. The resulting water molecule is therefore called polar. The molecules will tend to form a weak structure within the water, with the

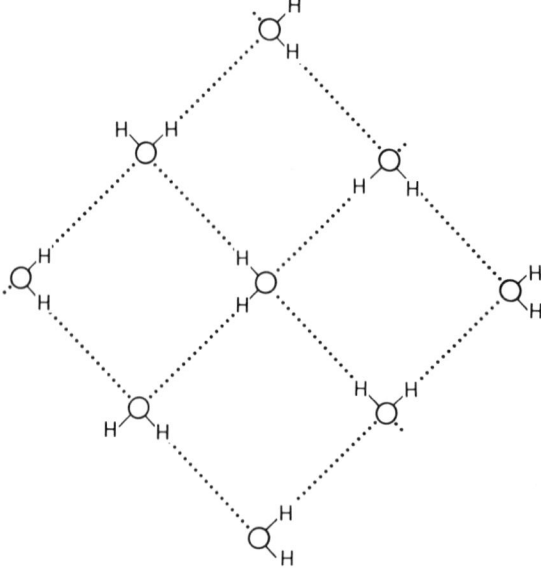

Figure 1.2 The polar nature of water means that it forms transient loose hydrogen bonded lattices when liquid

weak positive charges associating themselves with the weak negative charges of the neighbouring molecule, forming what is called a hydrogen bond. The effect is illustrated in *Figure 1.2*. It is this loose hydrogen bonding which gives the liquid water a more coherent structure which means that it does not behave as a predictable simple fluid. By all extrapolations, water should be a gas at room temperature. However, the bonding means that it has high values for viscosity, for surface tension and for latent heats.

Water can act as an acid or a base because it can be a proton donor or proton acceptor. It has great solvent power for ionic compounds and can readily dissolve many non-ionic covalent compounds because of its ability to form hydrogen bonds so readily. It is also colourless, tasteless and odourless.

Water plays a critical role in our lives in many subtle ways. It is a major component in ordinary air. It readily absorbs radiation and therefore its presence has a strong influence on the ability of infrared energy to penetrate the air. It adheres to most surfaces and in very thin molecular layers does not at first behave as water. As the molecular layers grow they become more water-like, and this then affects the electrical surface conductivity and can cause electrical breakdown and corrosion. Water is also essential for life because all organisms, from the simplest to the most complex, rely on aqueous solutions for metabolism.

In the following chapters the role of water vapour in our everyday life is reviewed.

Further reading

Anon. 'Water', in *The Encyclopaedia Britannica*, 15th edn. **19**, 633–637, University of Chicago, 1974
Dampier, W.C. *A History of Science.* Cambridge University Press, Cambridge, 1979
Dorsay, N.E. *Properties of Ordinary Water Substance in All its Phases.* Reinhold Publishing Corporation, New York, 1940
Eisenberg, D. and Kauzmann, W. *The Structure and Properties of Water.* Oxford University Press, Oxford, 1969
Kirk, R. E. and Othmer, D. F. *Kirk–Othmer Encyclopaedia of Chemical Technology*, Vol. 21. Interscience, New York, 1970
Robinson, R.A. and Stokes, R.H. *Electrolyte Solutions*, 2nd edition. Butterworths, London, 1959
Silcocks, C.G. *Physical Chemistry.* Macdonald & Evans, Plymouth, 1980

Chapter 2

Human perception of moisture

2.1 Introduction

The influence of moisture on people is one of the least understood of all the environmental factors. Research has been controversial and there are wide divergences of opinion amongst consultants and building operators[1-3]. There are five reasons for this:

(1) There are wide differences in sensitivity to moisture between people. Not only is this seen from their subjective responses[4,5] but it can also be measured in physiological terms of mucus flow on the throat[6]. This sensitivity may not be displayed fully in laboratory tests, which often use large numbers of students who are at their peak of physical fitness and health.

(2) The critical factor may be either relative humidity or absolute humidity. Tactile perception of fabrics[7] and the amount of electrostatic energy generated by walking on a carpet[8] are both closely linked to relative humidity. Moisture loss from the body is a function of the absolute humidity or vapour pressure[9].

(3) The time taken to notice the effects of changes in humidity is a matter of hours, not minutes. This has been reported by long-haul pilots who spend much of their working life in very low humidity conditions[10,11].

(4) Moisture can have indirect influences on people. For example, low relative humidities are associated with dustier atmospheres than those at high relative humidities. The dust burden may create some discomfort and enhance any effects of a dry throat which the dry conditions have created[2].

(5) Occasionally, particularly in research studies where people are moved from one environmental chamber to another and are then asked to assess the difference, the change in sensation may be an artefact. Transient moisture adsorption or desorption occurs in clothing when moving to a higher or lower relative humidity. This condensation or evaporation phenomenon produces a transient warm glow or chill. The effect is quite significant and real but is a function of the amount and type of clothing rather than a direct human response to the moisture.

There are at least eight types of influence which moisture can exert on people. These are:

(1) *Warmth, comfort and stress.* The evaporative cooling effect of insensible perspiration and respiration is slightly less at higher humidities. Once sweating occurs then the ambient humidity becomes very important and high humidities can create oppressive thermal discomfort. There are also other dimensions of comfort other than warmth, such as skin wettedness.

(2) *Dry throats and noses.* The nose acts as a powerful humidifier for the inhaled air. At low ambient humidities the moisture-release mechanism may be overwhelmed and the nose and throat become dry. This can become uncomfortable and lower the resistance to infection.

(3) *Eye comfort.* The lubricant for the eyeball is tear fluid surrounded by a thin oil film. In very dry conditions the evaporation rate can be abnormally high, and the eye can require much more frequent refreshing through blinking. If the work task demands intense concentration this may temporarily inhibit frequent blinking and may lead to a breakdown in the tear film with a consequent dry and itchy eye.

(4) *Skin comfort.* Skin requires a minimum water content to remain supple. If it becomes too dry it roughens and can crack. If it becomes too moist because perspiration is inhibited then clothing comfort deteriorates due to it sticking to the damp skin.

(5) *Clothing and fabrics.* The flexibility, permeability to perspiration, tactile dampness and thermal insulation properties of fabrics are strongly influenced by moisture.

(6) *Electrostatic shocks.* The electrical properties of fabrics are also a function of moisture content. The low moisture contents of carpets can induce electrostatic charges to people walking on them and create electrostatic shocks when the person touches an earthed conductor.

(7) *Ill health and allergies.* Damp conditions enable moulds and house mites to flourish. The spores from the moulds and the metabolites from the mites can activate allergic responses in susceptible occupants. On the other hand, very dry conditions are believed to lower the resistance to infection and lead to respiratory infection.

(8) *Pollution.* The influence of humidity on pollution is twofold. First, it can affect the pollution concentration. Dust burdens are more likely to be higher in dry atmospheres. The release of formaldehyde gas from certain building materials is sensitive to changes in ambient humidity[12]. Second, humidity can influence the personal sensation caused by the contaminant. Odours and irritant effects are usually perceived worse at low humidities.

Let us now examine these factors in turn.

2.2 Warmth, comfort and stress

The metabolic rate of a person is determined by two factors. The first is the basal rate which is needed to support life. This is typically 90 W for an adult male. The second is related to the amount of muscular external work being undertaken[9]. For light secretarial duties, for example, this would add a further 50 W to the basal rate. Women have a metabolic rate approximately 10% less than that of men due to their smaller physical size. This total metabolic energy has to be dissipated while maintaining a body core temperature of 37°C.

Some 70–80% of this metabolic heat is lost through radiation and convection to the room (*Figure 2.1*)[13]. The remaining 20–30% is lost through evaporation. In non-sweating conditions this amounts to approximately 20 g/hour by insensible perspiration through the skin, and another 20 g/hour through evaporation from the lungs.

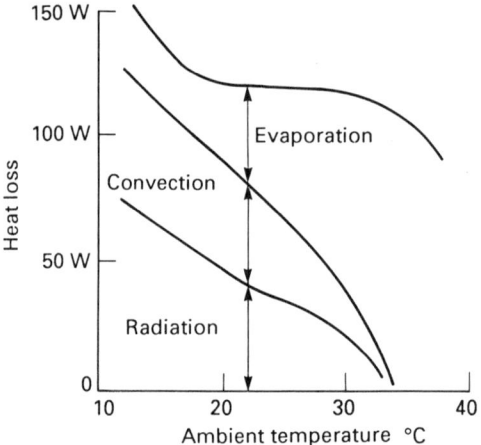

Figure 2.1 Heat-loss components for a clothed sedentary person

The insensible perspiration from the skin will vary with changes in the water vapour pressure in the room, diminishing with higher vapour pressures. Detailed physiological studies confirm this. An experiment at 19°C found that the insensible perspiration was 28 g/hour at 17% relative humidity (r.h.) and fell to 16 g/hour at 75% r.h.[14]. Most investigations have shown that for non-sweating conditions there will be a small increase in skin temperature and warmth with increasing humidity[14–17]. It requires an increase of 50% r.h. to increase forehead temperature by 0.3°C[15]. The results from a most comprehensive laboratory survey using 360 women and 360 men aged between 18 and 23 years and wearing standardized clothing of thermal insulation 0.52 clo (clo is a unit of insulation; see later) are illustrated in *Figure 2.2.*[18]. This shows that it requires an increase of 30% r.h. to produce the same warmth as a 1°C rise in temperature. The results also show that there are different responses between men and women at high humidities. The warmth effect of high relative humidities is more pronounced for men.

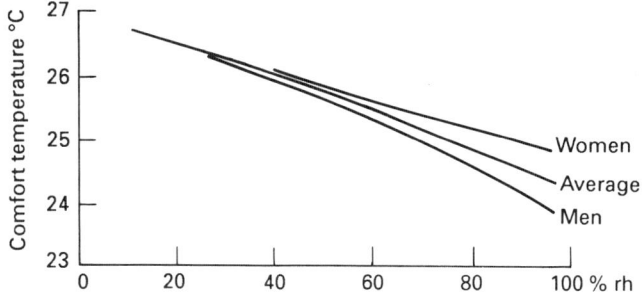

Figure 2.2 The influence of relative humidity on warmth

Office studies show that reports of insufficient moisture are commonplace, particularly in winter[3,5], even though the actual relative humidity measurements lie between 40 and 65%. This is reinforced by a laboratory study which showed that 48 males dressed uniformly in cotton garments with an insulating value of 0.7 clo regarded 70% r.h. at 23°C as their favoured relative humidity[2]. Further studies suggest that there may be an optimum relative humidity for each temperature because air temperature and ambient water vapour pressure exert opposing influences on the rate of evaporation of moisture from the mucous membranes. Raised humidity at 23–24°C caused 70% of the occupants to judge the air 'stuffy' compared with only 10% at 21–23°C[3].

Field studies also reveal that twice as many women complain of dryness than men when working in low relative humidities. Under these dry conditions there are also twice the number of complaints from those over 35 years old than from the younger part of the working population. The demand for more humidification increased with the time spent working in the office. Dissatisfaction with dry conditions was five times more from those who had worked in the office for at least three years compared with newcomers[3].

It is interesting to note that the field surveys treated moisture as an independent factor unrelated to warmth sensation. Early studies[15] suggested that skin dampness became uncomfortable above relative humidities of 70%, and others have since proposed that the degree of skin wettedness should be considered as a possible comfort factor. This aspect is discussed further in Chapter 4[19,20].

German experience suggests that an oppressive sultriness occurs at a water vapour pressure of 1.9 kPa. This is attributed to the onset of sweating and therefore an upper limit slightly below this value has been adopted in an approximate way in their national standard[13,21]. This is applicable to sedentary and clothed adults. This comfort line is shown in *Figure 2.3*. Other researchers show similar results[22-24].

The overall conclusion for a non-sweating condition is that humidity will have an effect but it will be small. Work conditions which are modest in temperature around 21°C and high in relative humidity (65–70%) appear to be the most attractive.

Heat stress occurs when the body is perspiring to overcome excessive heat gain. This stress has been quantified in three ways. The simplest was

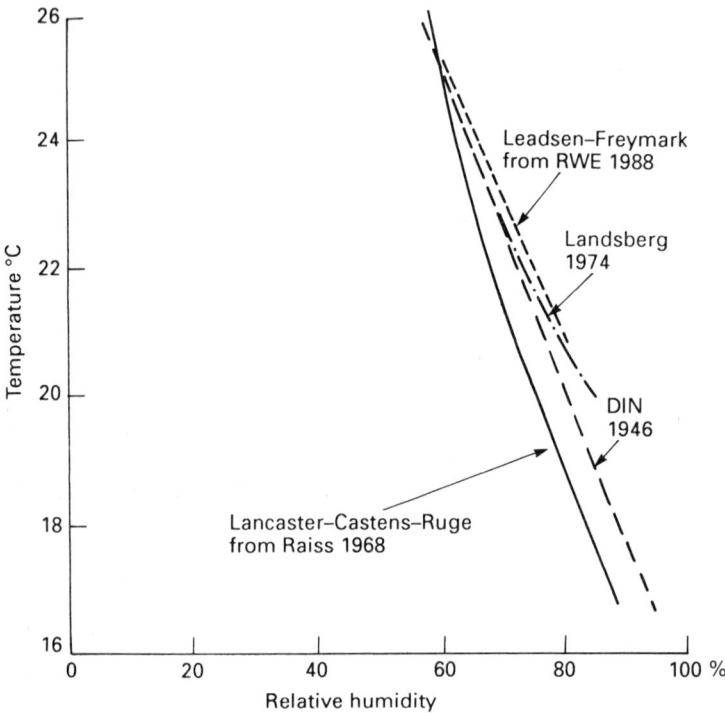

Figure 2.3 Upper limit of relative humidity to avoid sultriness

devised by the US Weather Bureau and represented on a temperature scale the discomfort which humidity and air temperature could create on a warm humid day[25]. This Temperature–Humidity Index is represented as

$$\text{THI} = 0.4 \, (T_{\text{air}} + T_{\text{wet bulb}}) + 4.8$$

where the temperatures are measured in degrees Celsius.

The index represented modest stress accurately but made no allowance for air speed or thermal radiation.

A more comprehensive approach to the strain produced by heat stress is a nominal value which represents the amount of sweat secreted by fit, acclimatized young men in four hours[26]. This index allows for the influences of the environmental factors, air temperature, mean radiant temperature, air speed and humidity, and the personal ones of metabolic rate and clothing insulation. The index is empirical, based on laboratory measurements.

In very special cases of high temperature and high humidity, such as that of mine rescue workers, the evaporative cooling ability of the person can be hampered either by high humidity limiting the rate of evaporation of sweat or because the body cannot secrete sufficient sweat to provide the required cooling. The body temperature will then slowly rise. The safety criterion is then the safe upper body temperature for the person. The stress index will assess, for any given conditions, the time needed to reach this critical temperature. The recommended maximum body temperature is 38°C.

An illustration of practical guidelines on permissible working time in different conditions is given in *Figure 2.4*[13].

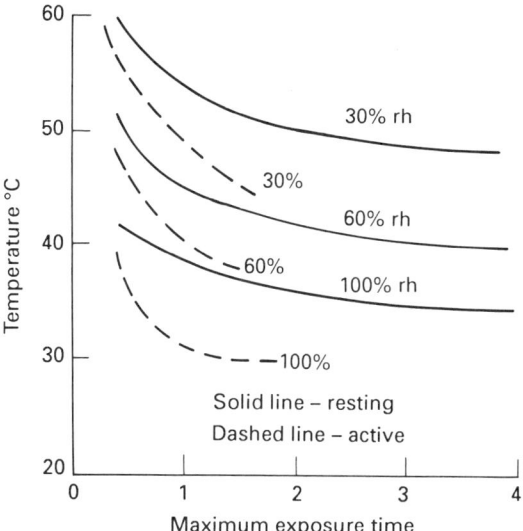

Figure 2.4 Maximum exposure to heat stress

2.3 Dry throats and noses

The major human respiratory passages are lined with a mucus layer which both moisturizes the air inhaled and simultaneously traps germs and particles. This mucus is secreted by minute glands at the surface of the epithelium. The dust-laden mucus is constantly driven towards the mouth by a carpet of fine hairs. These hairs flick the mucus upwards at a speed of approximately 5 mm/min.

If the mucus loses moisture it will become more viscous and would be expected to move more slowly and in extreme cases dry up completely. This dryness is noticeable in the nose and throat at low humidities and leads to discomfort. Patients have reported dry noses when the weather is sufficiently dry to make the indoor relative humidity fall to 25% r.h.[27].

Research studies are not in agreement on the conditions under which dry noses and throats occur. One early study of the back of the throat showed that it became dry when the water vapour pressure fell below 1.4 kPa. This is equivalent to 55% r.h. at 20°C[85]. A separate study of the mucus flow within the noses of 165 people showed a very clear relationship between ambient relative humidity and mucus flow. Raising the relative humidity from 30% to 70% r.h. doubled the flow (*Figure 2.5*)[6]. Smokers had a lower flow rate than non-smokers. The variation between people was particularly wide at the low relative humidities. The topic is still controversial[2, 28].

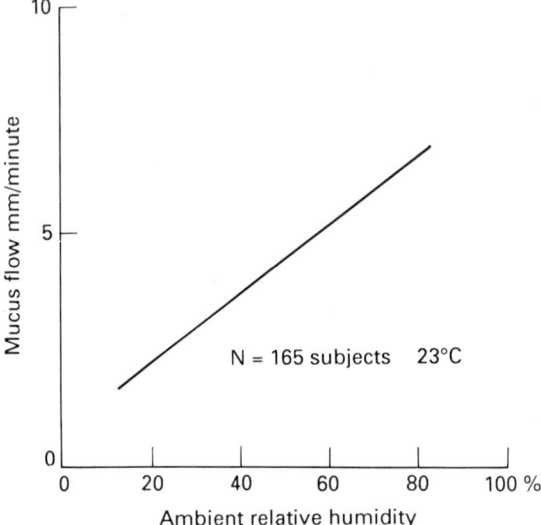

Figure 2.5 Mean rates of mucus flow in the nose

Scandinavian studies on aircraft suggest that it takes 2–3 hours for very low humidities (10% r.h.) to cause discomfort. Pilots do not experience dryness troubles for flights of less than two hours. However, 80% of pilots on flights of 3–8 hours duration complained of dryness in the nose, mouth and throat, often accompanied by itching in the nose and eyes. They also reported dry lips and dry skin[29].

Experiments on simulated space flight experience in the laboratory confirmed the commercial pilots' experiences. Dry conditions brought about complaints of nasal stuffiness, dry lips and dry throat. These effects were observed after approximately two hours exposure to 9% r.h. at 24°C. It was also supported by anecdotal evidence from long-distance jet pilots on repeated trans-Pacific journeys[11].

Dryness in the throat and nose is clearly linked to the ambient water vapour pressure. All the research shows wide variability between individuals at low humidities. The evidence suggests that for some the mucus will start to dry out below 1.4 kPa water vapour pressure and that 80% will be reporting dryness at 0.3 kPa (10% r.h. at 24°C).

2.4 Eye comfort

For the eye to be comfortable it needs to be covered with a thin fluid film. This fluid system serves at least four functions[30–33].

(1) Good optical surface to the cornea. Almost all of the refractive power of the eye originates at the outer surface of the cornea. The smooth optical boundary presented by the liquid surface provides the highest visual quality.

(2) A lubricant to the eyelids.

(3) Bacterial protection provided by traces of the enzymes lysozyme and beta-lysin within the tear fluid.

(4) An irrigation system which, particularly when irritated, will flood unwanted debris or noxious chemicals out of the eye.

The general system is illustrated in *Figure 2.6*. The aqueous tears are supplied by lacrimal glands sited in the outer region of the upper eyelids. The tears flow through ducts to the upper lid around the meniscus to the lower lid and drain through a lacrimal duct into the nose. The eyelid wipes the eyeball whenever a blink occurs, which is approximately every three seconds.

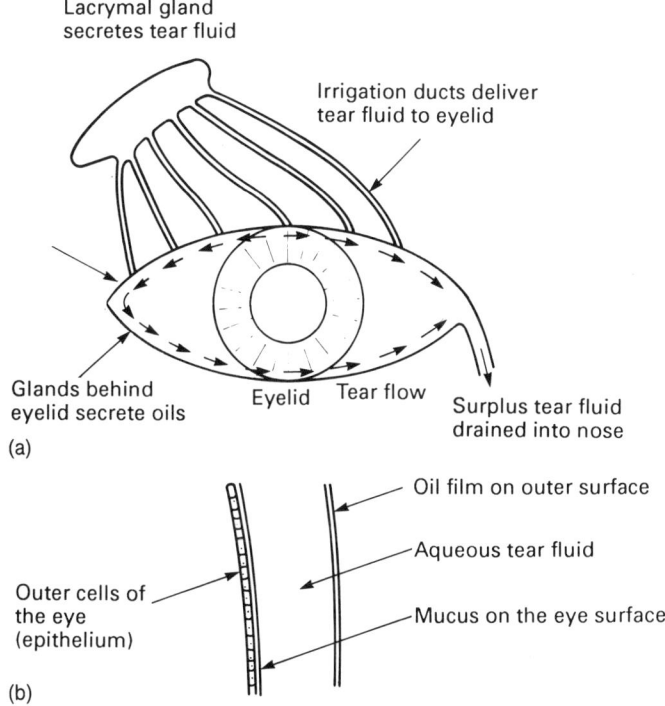

Figure 2.6 Fluid system for the eye. (a) Irrigation system of the eye. (b) Details of water film on the front of the eye (cornea)

The physical chemistry of the tear film is complex. There are three components. These are illustrated in *Figure 2.6*. The outermost layer is an oily film $0.1\,\mu m$ thick, made of lipids such as cholesteryl esters, supplied from meibomian glands in the eyelid. The base layer on the outer surface of the eyeball is mucus approximately $0.02–0.05\,\mu m$ thick. All three components are essential. The oily outer film inhibits evaporation, the water is the bulk fluid and the mucus layer makes the eyeball readily wettable[34–36].

In comfortable circumstances with a normal blink every three seconds or so the exposed part of the eyeball stays covered in the thin film. The film thickness starts at approximately 10 μm and evaporation thins it to 4 μm before the next blink[37]. If between blinks the evaporation occurs too quickly then dry spots appear on the cornea. These start as tiny dry holes which quickly grow and join with others to create a dry eye. Dry eyes are uncomfortable and even painful and produce sensations of burning and scratching. They can eventually damage the eye.

Laboratory measurements show a very wide variation between people in the time for a normal eye to become dry and this can vary from 10 to over 45 seconds with the bulk being between 15 and 34 seconds[38]. Each individual is consistent and both eyes in any one person are similar.

The symptoms of dry eyes occur in low-humidity conditions and in places of high air movement. Both of these factors enhance evaporation from the eye. Aircraft often operate at 10% r.h., offices in wintertime can operate around 20–30% and cars have ventilation systems which are designed to blow air on the windscreen, and high velocities can occur around the heads of the driver and passenger.

The very wide variability between people means that some are much more susceptible to dry eye than others. The time for an eye to become dry is termed break-up time (BUT) of the fluid film. Those individuals who have a short film BUT are much more susceptible to dry office conditions. In conventional offices the level of eye complaints is quite high. One Danish study found that 25% of the office workers experienced eye irritation several times per week. This sensitive group were shown to have shorter film BUTs than those who did not experience eye irritation[39]. One laboratory study showed a marked increase in eye discomfort when the ambient relative humidity went down to 20% r.h.[40].

The normal way of assessing the time for the cornea to go dry is to use a skilled ophthalmologist to watch the surface of the eye carefully through a microscope in special lighting conditions. A new way is to ask the individual to avoid blinking until he feels that his eyes have gone dry. This technique has been compared with the more objective opthalmic appraisal and found to be closely related, although the reported times to dryness are considerably longer. The eyes actually went dry in 13 seconds, although the individual involved believed it to be 46 seconds[41].

Contact lens wearers also experience the dry eye problem, although one survey suggests that contact lenses offer some protection to the eye, at least for the first few hours in dry conditions[42]. The cornea has no blood circulation and derives part of its oxygen from the atmosphere. If air is excluded, the lactic acid concentration rises owing to anaerobic metabolism. Extended wear contact lenses are made from a hydrogel which retains a very high water content and enables gaseous diffusion to occur more freely to and from the eye.

One office study on dry eyes with contact lenses measured the film BUT at the individual's place of work and recorded the local indoor relative humidity. On a discomfort scale of 0–3 where 0 = no discomfort, the office workers wearing contact lenses could be divided into those with discomfort and those without. Approximately one-third had no discomfort. The average vote of discomfort for the remaining two-thirds was 1.7. The relative humidity at the places of work for those with discomfort was lower than for those without discomfort, and the discomfort rose with reducing relative humidity. The film BUT was shorter for those experiencing eye discomfort. Those working in relative humidities below 31% also experienced more lens deposits[43]. These results are shown in *Figure 2.7*.

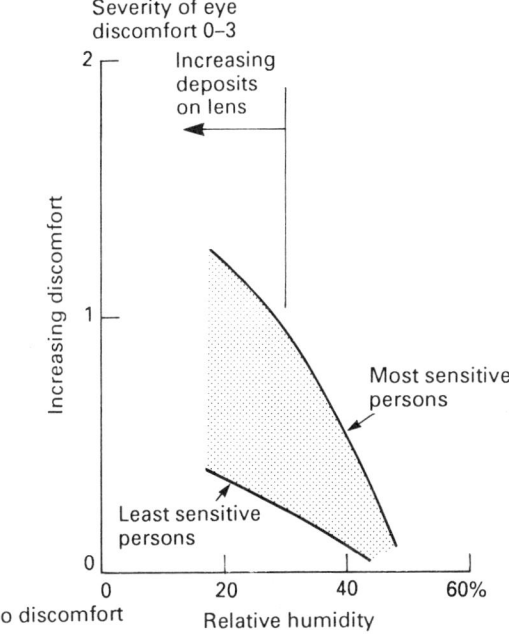

Figure 2.7 The influence of relative humidity on eye discomfort with contact lenses

A special experiment using five subjects with very short film BUTs and five with longer ones explored the effect of changing the relative humidity in the room and recording their discomfort while wearing contact lenses. The short BUT film people experienced more distress and it occurred more quickly than in the other group. Discomfort occurred within 15 minutes. The long BUT group experienced much less discomfort and it took an hour to reach their maximum discomfort. Relative humidities above 50% appeared to satisfy both groups.

Laboratory experiments have examined how people wearing contact lenses respond to low relative humidities. The experiments used one eye with a contact lens and the other without. The questionnaire appraisal technique revealed two types of response. The first was a dimension of pain represented by scratchiness, tearfulness and a burning sensation. The second was annoyance represented by discomfort, tiredness, strain, and moisture perception. The annoyance scores rose rapidly in the first hour of the test at low relative humidities (20% and 30% r.h.) and continued to rise for the 10 hours of the test. The scores representing painfulness increased only slowly during the day. This effect is illustrated in *Figure 2.8*[42].

Studies on the contact lenses themselves show how their water content does vary with ambient relative humidity[44, 45].

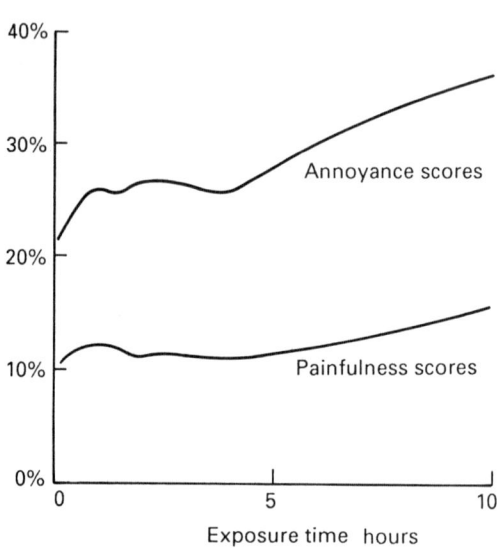

Figure 2.8 Mean eye discomfort scores change with time in low-humidity conditions

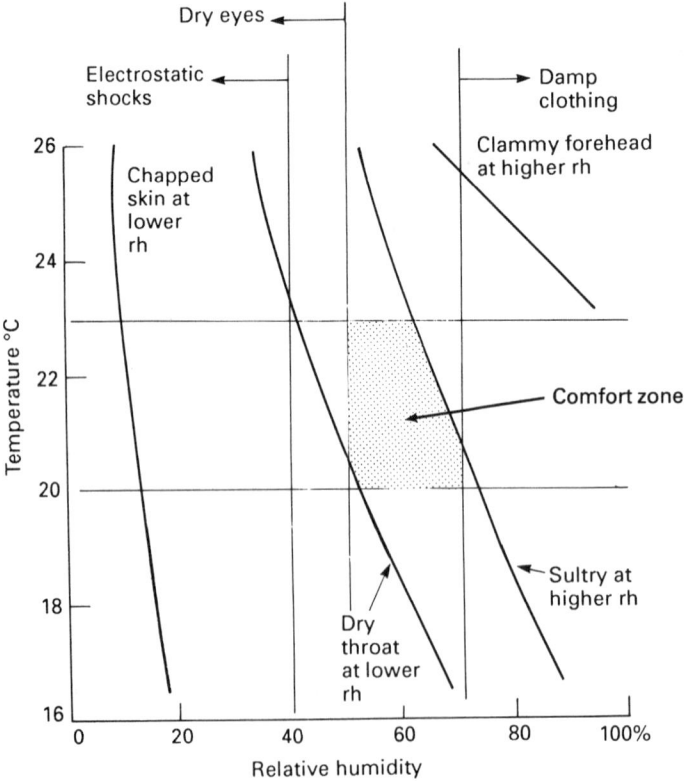

Figure 2.9 The comfort envelope

Studies on eye dryness are relatively new and it is only in the last 30 years that the physical mechanism of the fluid layer has been understood in its subtle complexity. The four main findings from research to date are:

(1) There is a very wide difference between individuals in the time taken for the cornea of the eye to go dry in any given environment. There is normally a big safety margin between the average duration between blinks which is 3 seconds and the time needed for the liquid film on the eye to dry out, which takes on average 13 seconds.

(2) Those individuals who have the shortest time for the eye to dry out are more likely to experience annc ice and discomfort in either dry atmospheres or in atmospheres with air movement around the eyes. Simple self-diagnostic tests are being developed so that susceptible individuals can identify themselves.

(3) The distress in dry atmospheres manifests itself progressively in 15 minutes for sensitive individuals but can take up to one hour for others. The distress continues to increase slowly with exposure.

(4) Preliminary findings suggest that relative humidities above 50% provide comfort conditions for the eye.

The lower limit of relative humidity of 50% is plotted on the comfort envelope illustrated in *Figure 2.9*.

2.5 Skin comfort

An adult is clad in $1.8\,m^2$ of skin. The toughness, thickness, elasticity and flexibility match the special needs of particular parts of the body. This skin weighs 3–4 kg and therefore forms one of the major organs of the body[46].

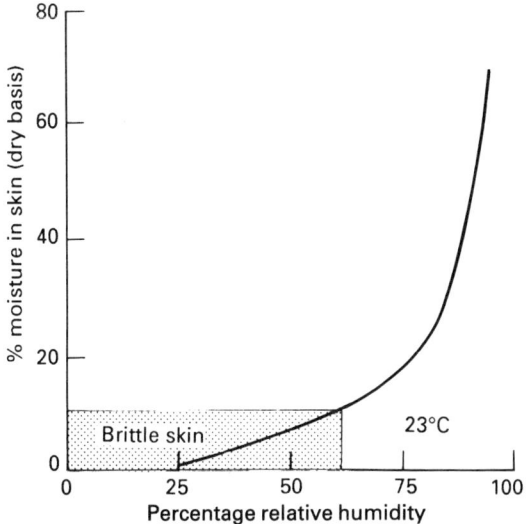

Figure 2.10 Adsorption isotherm for excised skin

It serves far more than as a self-repairing packaging material because it provides the vehicle for tactile sensation, pain and thermal sensations. In addition, it supplies protection from ultraviolet irradiation and contains three million sweat glands for evaporative cooling in heat stress. These sweat glands are particularly profuse on the hands and soles of the feet to keep the skin moist and hence flexible and able to grip surfaces. It also provides variable thermal insulation to protect against cold.

Exploratory experiments on excised skin in the laboratory show it to be very sensitive to moisture. The adsorption isotherm is very hygroscopic (*Figure 2.10*). The physical characteristics of skin change when the moisture content falls below 10%, when the skin becomes brittle and looks dry[47]. This moisture content is reached when the ambient relative humidity reaches 60%. In practice, body moisture is constantly permeating the skin and therefore the moisture content of skin will be above the critical 10% value at much lower ambient relative humidities. Once excised skin is above this 10% moisture content it becomes supple and its elasticity increases ten-fold in going from equilibrium in an atmosphere at 65% r.h. to one at 100% r.h.[48]. This dramatic change is illustrated in *Figure 2.11*.

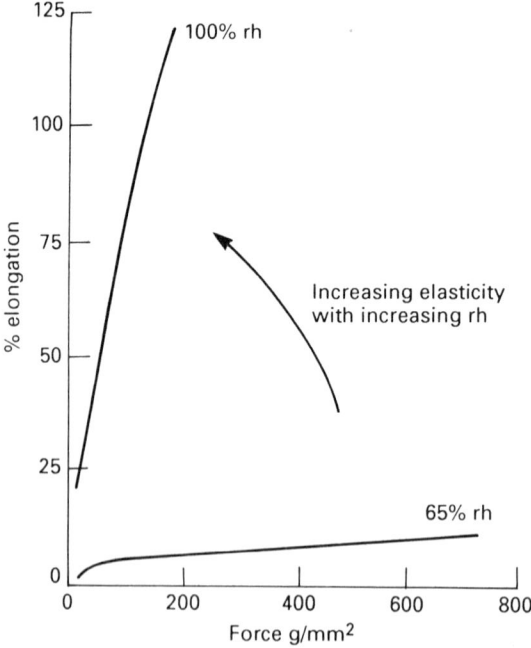

Figure 2.11 Excised skin elasticity becomes greater for atmospheres above 65% r.h.

Experiments in four small skin sites on the forearms of 250 people showed that skin can either release or absorb water vapour from the atmosphere. Each individual had a critical relative humidity where the skin neither lost nor gained moisture. This critical relative humidity was unaffected by exposure times from 30 minutes to 8 hours, by season of the year, and, surprisingly, by skin temperature, excluding sweating[49]. The results showed a very wide difference between individuals. The average of all tests was a critical relative humidity of 86% but three clusters of results occurred. The low relative humidity group had a critical relative humidity around 60–70% and tended to be subjects experiencing poor health. The normal group had a critical relative humidity of 75–82%. A high group around r.h. 85–90% were often based on values when sweating occurred. Higher humidities than the critical values would lead to higher skin moisture content and could lead to uncomfortable and sticky clothes.

In very dry ambient conditions body skin does go below this critical moisture content. The first effect is that the dead flattened skin cells which form the outermost layer of skin lose their cohesion and the skin surface becomes rough. This condition can occur after a few hours exposure to a very dry atmosphere and can disappear as quickly on returning to more humid conditions. If the dryness is intense then the skin can become chapped and cracked and if the basal layer of growing cells is torn then the skin fissures will be slow to heal. This is illustrated in *Figure 2.12*[50].

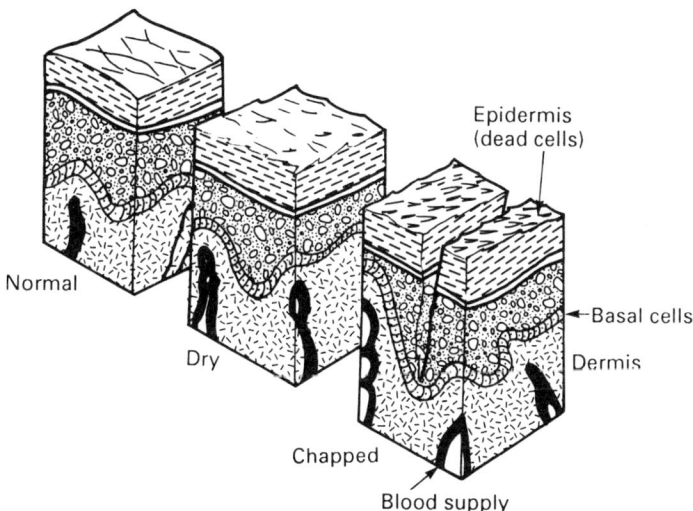

Figure 2.12 Progressive effect of dry skin conditions

Medical evidence suggests that dry skin problems are worse with people in those occupations which are associated with intermittent wetting of the skin. Children are more susceptible than adults. Some people are affected much more readily than others and this sensitivity is hereditary. Field studies associate skin problems with outdoor dew points below −7°C, which is equivalent to 15% r.h. at 20°C[51]. The effect becomes progressively more severe at even lower dew points with associated skin fissures[52]. The phenomenon is uncommon in temperate climates but well-known in colder climates.

Careful environmental laboratory assessments have shown that people do perceive skin moisture and that these perceptions are clearly linked to both temperature and relative humidity. The most comfortable relative

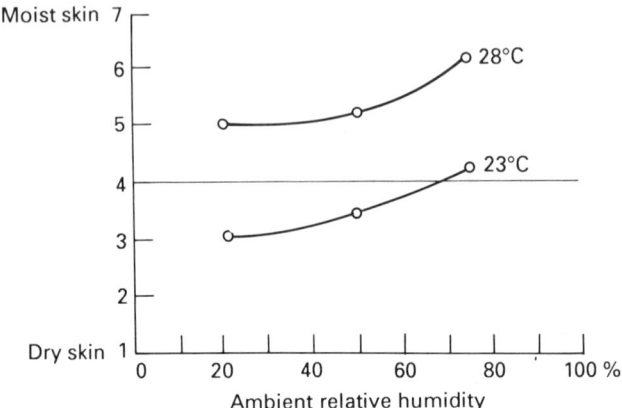

Figure 2.13 Perception of skin moisture varies with humidity

humidity for skin was 70% at 23°C. At 28°C room temperature, sweating was imminent and the skin was always too moist at all the relative humidites (20–80% r.h.). This relationship is illustrated in *Figure 2.13*[53].

Pleasant environments do not induce sweating. Early American studies noted that the onset of sweating was marked by a clammy skin sensation over the whole body, including the forehead. Further heat stress led to body dampness and beads of sweat on the forehead. While there are wide differences between individuals, the conditions under which at least 40% of the subjects experienced the onset of sweating are shown in *Figure 2.9*[54].

Laboratory studies inquiring into dryness and discomfort at low humidities found that the moisture loss from a person was very high at first and that it took 20 hours to adapt to it[11]. Dry lips were the first signs noticed but were not immediately reported as uncomfortable. The discomfort symptoms in order of diminishing importance were:

(1) Nose stuffiness
(2) Dry lips
(3) Dry throat
(4) Dry eyes
(5) Dry tongue
(6) Dry scalp
(7) Dry skin

The overall conclusion is that people can sensitively detect the moisture status of their skin and that the favoured condition is modest room temperature 21°C and a high relative humidity of 70%.

2.6 Clothing and fabrics

All fabrics take up moisture as the ambient humidity rises. The amount of moisture is determined by the relative humidity, not the water vapour pressure, and is only slightly affected by temperature. Organic fibres such as wool, cotton and linen adsorb large amounts of moisture, particularly at

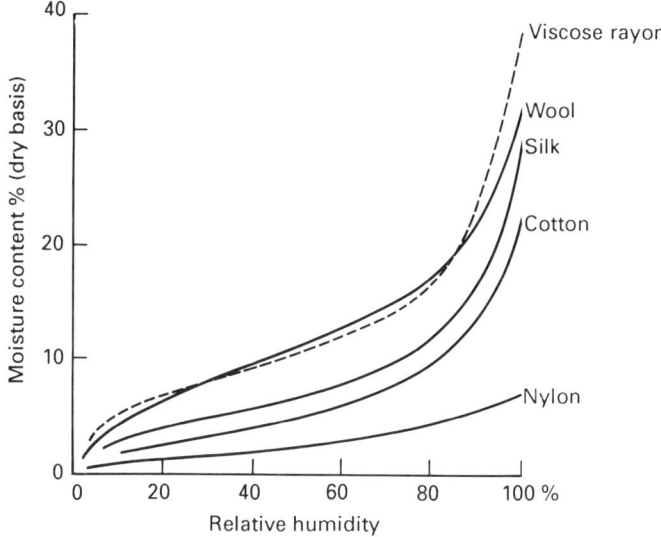

Figure 2.14 Adsorption isotherms for clothing materials

high relative humidities. Artificial fibres usually adsorb much less. Adsorption isotherms are shown in *Figure 2.14*[55].

The subjective chilling effect of cold fabrics placed upon the skin is determined by the degree of intimate contact the material makes with skin. Laboratory experiments have studied the temperature drop which occurs when fabrics are placed on hotplates designed to simulate warm skin. This temperature drop is negligible for the fabrics with a rough surface such as wool and ordinary cotton, small for smooth bleached cotton and highest for silk and acetate rayon[56,57]. Moisture effects were unimportant.

The feeling of dampness in a fabric is also influenced by its surface properties, but for each material it is clearly linked to moisture content. Surprisingly it is not closely linked with the actual moisture content but more with the adsorption isotherm. Perceptions of dampness for both cotton towels and woollen socks varied with ambient relative humidity in an identical manner, despite the woollen socks containing twice the water content of the cotton towels. The very wide range of people's perceptions of dampness is shown in *Figure 2.15*[7]. Polyester–cotton napkins were perceived differently, with a significant proportion of the population finding the fabric to be damp even at almost complete dryness. This unusual effect is attributed to the glazed feel of the napkins and its influence on 'chilliness' rather than 'dampness'. Polyester–cotton was considered dry by 50% of the respondents at 75% r.h., cotton towels and woollen socks at 50% r.h.

Comfortable clothing has three important physical characteristics. The first is thermal insulation. In general this is directly proportional to the thickness. The insulation unit is the clo, which represents an average business suit[58-80]. This property is now well-recognized and quantified for sedentary work. Research is exploring how activity provides a bellows pumping action. Moisture does have an effect on the thermal conductivity

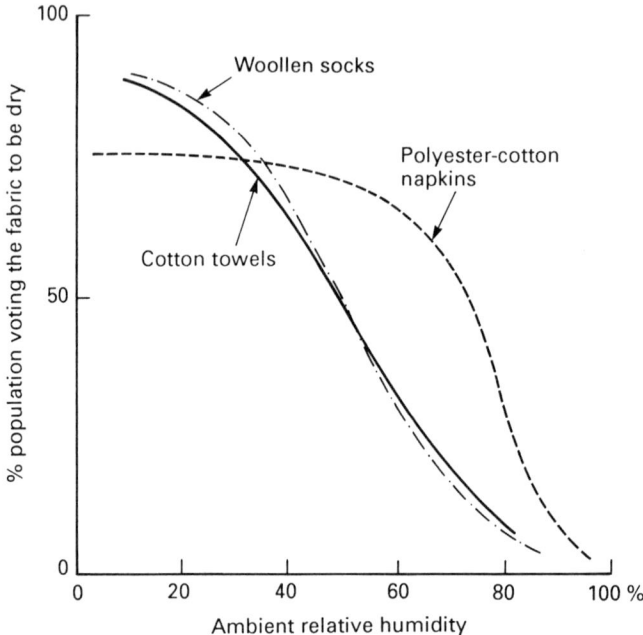

Figure 2.15 Perception of a dry fabric varies with relative humidity

of clothing, increasing it with increasing moisture content. The insulation value is particularly good at very low moisture contents. The research is limited but two studies show similar effects. The first investigated underclothing[61] and the second studied thermal conductivity in a laboratory simulation. The results from this experiment are illustrated in *Figure 2.16*[62].

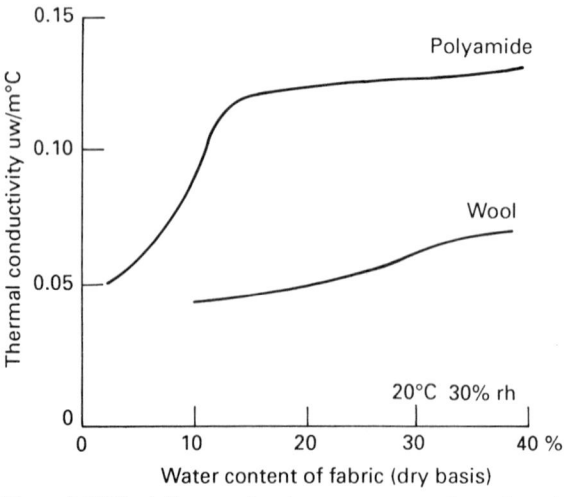

Figure 2.16 The influence of moisture content on thermal conductivity

The second effect is moisture permeability. Permeability is necessary in clothing to allow the insensible perspiration through the skin to dissipate itself. Clothing must permit this moisture vapour flow. This permeability is a function of the clothing weave and the material itself. Diffusion resistance increases with higher moisture content. The third effect is the wicking of moisture or capillarity through the fabric. This is determined by the wettability of the fabric which in turn is strongly influenced by any surface finish. An illustration of variation in these two factors with moisture content is shown in *Figure 2.17*[62].

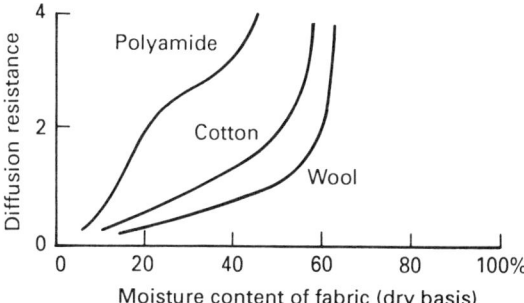

(a) Permeability

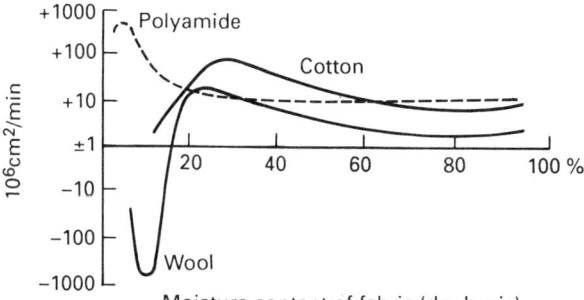

(b) Capillarity

Figure 2.17 Moisture content affects both the permeability and capillarity or wicking of the fabric

The relative importance of the three factors of insulation, permeability and capillarity varies with different climates. In very cold climates insulation and permeability are critical. In hot humid climates the permeability and capillarity are important. Capillarity determines the upper comfort limit for humid climates. Linen performs best in these circumstances and can be worn in atmospheres up to 96% r.h. Wool and cotton are similar and are comfortable up to 80% r.h. Artificial fibres are satisfactory up to 70% r.h.

Dampness also affects the compressibility of clothing, as many fibres lose their natural springiness in moist conditions. Compressed clothing is not as good a thermal insulator[63].

2.7 Electrostatic shocks

A person walking across a floor generates a small amount of electric charge. If the floor is electrically conducting then this charge is quickly dissipated to earth. If the floor is electrically insulating then the charge cannot leak away and the voltage of the person rises with each step until many minor corona discharges occur which limit the upper voltage. This phenomenon is illustrated in *Figure 2.18*. Whenever the charged person then touches an earthed object such as a door handle, the stored electrical energy is discharged through a spark which creates a pin-sized burn on the hand, termed an electrostatic shock.

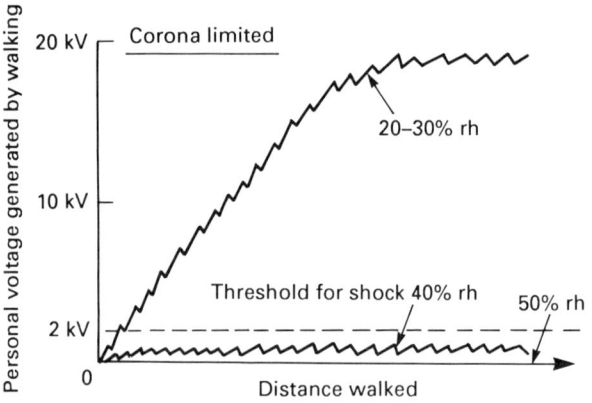

Figure 2.18 Electrostatic shocks are related to ambient relative humidity

The voltage generated is a direct function of the electrical resistance of the floor and of the distance walked. The electrical resistance of waxed floors and of carpets depends upon the ambient relative humidity. Most floors have sufficient electrical insulation below 30% r.h. to create electrostatic problems. In practice the electrical insulation is a complex function of the carpet weave, the underlay and the component materials[8]. Some carpets give electrostatic problems even at 40% r.h. but seldom at any higher relative humidity.

Underfloor heating poses special problems because the carpet temperature is higher than normal. This means that the relative humidity within the carpet will be lower than that in the room by approximately 20%. The room humidity will have to be higher than normal to avoid electrostatic shocks, i.e. around 55% r.h. or higher.

There is a wide variation in sensitivity to electrostatic shocks between different people but in general women are more sensitive than men.

Much more detailed information is given in Chapter 7 on electrostatics.

2.8 Air pollution

Humidity has two different types of influence on air pollution. It can affect the generation of pollutants and it can also modify the perception of the pollution.

Strong seasonal changes in outdoor dust concentrations have coincided with the normal dry spells in Germany at the end of summer and early autumn[64]. Experience has shown that dust burdens within buildings increase particularly at values of relative humidity below 35%[13]. Studies of office dust show it to be fine fibrous cellulose from paper products blended with powdered materials which could be scales of dry skin, together with traces of soil and clay from the neighbouring ground[65].

The dried tobacco leaf used in cigarettes is hydroscopic and therefore takes up moisture when placed in conditions of higher relative humidity. High-moisture tobaccos burn more slowly and generate less smoke and more odour. This phenomenon has been quantified for an American tobacco and is illustrated in *Figure 2.19*[66]. In general, more tobacco is burned during the smoulder period than during the puff period[67]. The smoke emitted during this smoulder period is called the sidestream smoke. The nicotine content of this smoke increases with higher moisture content tobacco[68, 69].

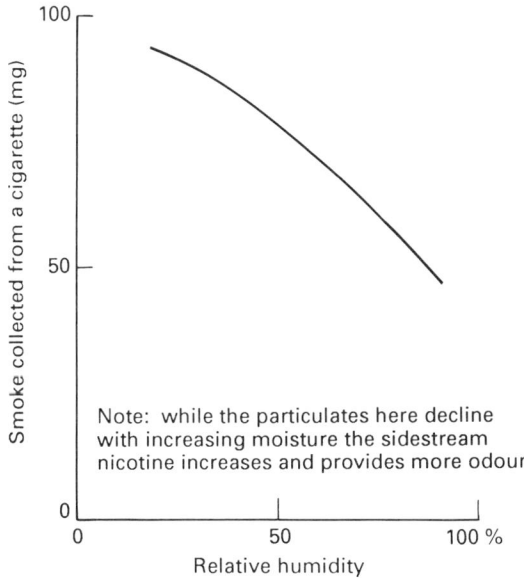

Note: while the particulates here decline with increasing moisture the sidestream nicotine increases and provides more odour

Figure 2.19 The influence of relative humidity on cigarette smoke

Simulated aircraft studies maintained a constant smoke concentration inside the space and investigated the occupants' response to it. Odour was not a problem because the occupants rapidly adapted to it. The discomfort criteria were all irritations and these reached maximum response after 25–30 minutes exposure. Eyes, nose and throat were affected. Personal differences in reactions to the same smoke concentration were particularly high[70]. This irritation is influenced by both temperature and humidity. General irritation with poor air quality is the most sensitive factor, with non-smokers registering protests at much lower smoke concentrations than the smokers. Eye irritation was the next most sensitive factor, with little difference in response between smokers and non-smokers. Nose irritation was less for smokers. Two relative humidities, 33% and 85%, were used for the tests and all were at 25°C. The worst irritation occurred at the lower relative humidity[71]. However, subsequent tests in cooler conditions (18–19°C) showed a smaller influence of relative humidity with the worst irritation reported at moderate relative humidities[72].

The influence of temperature and humidity on the odour perception of cigarette smoke has been studied in the laboratory. Assessments were made by a panel of trained staff divided equally between men and women and moderate smokers and non-smokers. The technique used was the very sensitive one of walking into the room and sniffing for a first impression and repeating the sniff five seconds later. The odour intensity was diminished by both increasing temperature and by increasing humidity. In the comfort region of 21°C the change in relative humidity from 30% to 60% lowered the odour intensity by one-quarter of a vote. This is approximately the same reduction which a rise of 5°C would create. The results of this study are illustrated in *Figure 2.20*. Irritation had a similar pattern but was not so clearly defined[73].

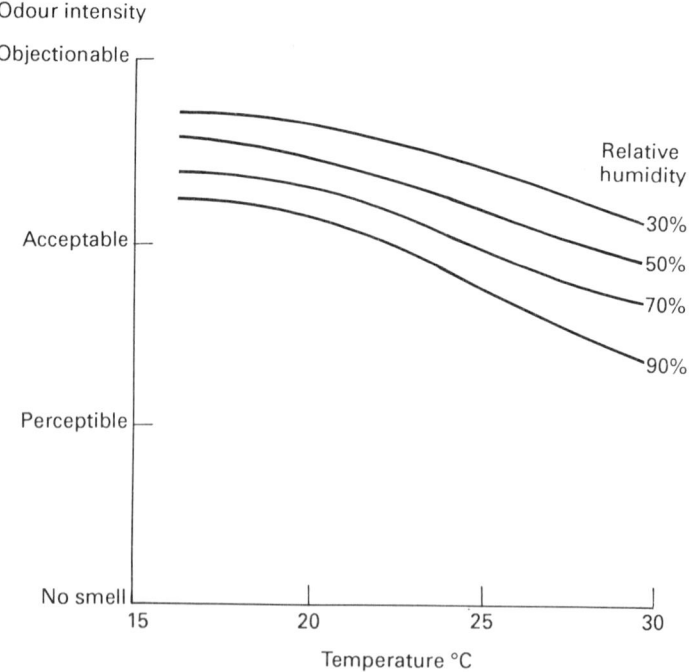

Figure 2.20 The influence of temperature and humidity on odour intensity of cigarette smoke

The speed of adaptation to odour and the steady rise of irritation was studied over a six-minute exposure for three relative humidities. These results are shown in *Figure 2.21*. The rapid adaptation to the odour over the first two minutes was matched by a corresponding increase in irritation over the same period.

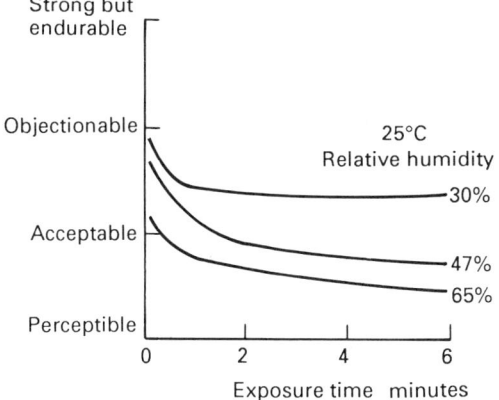

(a) Odour perception decreases with exposure time

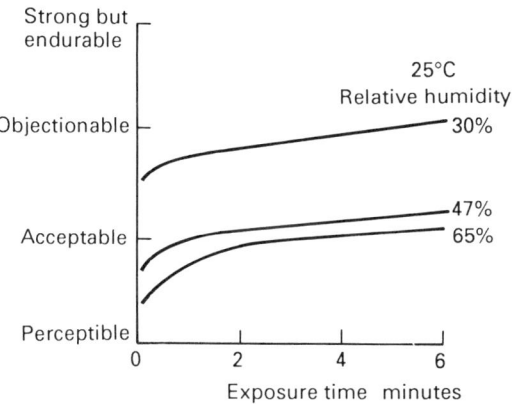

(b) Irritation increases with exposure time

Figure 2.21 Odour and irritation caused by cigarette smoke change with time and relative humidity

In countries where chipboard is widely used as a building construction material there can be small quantities of unpleasant formaldehyde gas released into the room. This comes from imperfectly cured binder present in the chipboard. Laboratory experiments have shown that the equilibrium concentration of formaldehyde was directly proportional to the ambient water vapour pressure and to the temperature[74].

2.9 Illness

Coughs and colds are associated with winter conditions. Popular beliefs attribute such illnesses to lack of sunshine, lack of exercise and perhaps a more limited diet, all aided by a better infection route through more time spent indoors, and overcrowding in poorly ventilated spaces. Studies on colds have shown consistently that it is the abnormally cold weather which is linked to rises in the incidence of upper respiratory infections[75, 76]. This could mean that indoor relative humidity may be a critical factor.

Some medical practitioners associate dry throat conditions with the onset of a cough or cold[77, 78]. Field studies have been controversial but some have shown a benefit by raising the relative humidity to 40%[79, 80]. An analysis of school absenteeism in 11 Saskatoon schools suggests a relationship with indoor relative humidity, although there is a lot of scatter in the raw data points. This is shown in *Figure 2.22*.

The problem appears to be more acute in very cold climates, where indoor relative humidities can fall to 20% in winter.

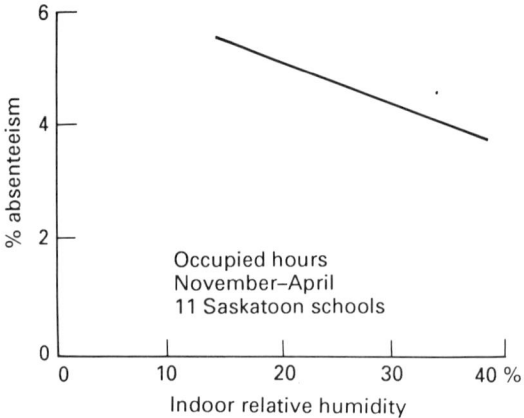

Figure 2.22 The relation between school absenteeism and indoor relative humidity

2.10 Comfort guides

Professional building service engineers prepare codes of practice for their own designs. One of the largest used internationally is that of the American Society of Heating and Refrigeration and Air Conditioning Engineers, who use thermal comfort charts which span 25–90% r.h. for a range of temperatures. However, a narrower comfort band of 45–60% r.h. is recommended to minimize the influence of odour and irritant pollutants such as cigarette smoke. The Chartered Institution of Building Service Engineers, Britain, recommends 40–70% r.h. The Swedish National Board of Industrial Safety (Arbetarskyddsstyrelsen 1967) recommends a minimum relative humidity of 30%. The German DIN recommendations link humidity with temperature in their highly developed approach to sultriness.

Other authorities quote slightly different conclusions, sometimes with associated temperatures. In general, the earlier references favoured the lower temperatures and higher relative humidities, while more recently temperatures have been rising and lower values of permissible relative humidity are being recommended.

These guides are summarized in *Figure 2.23*.

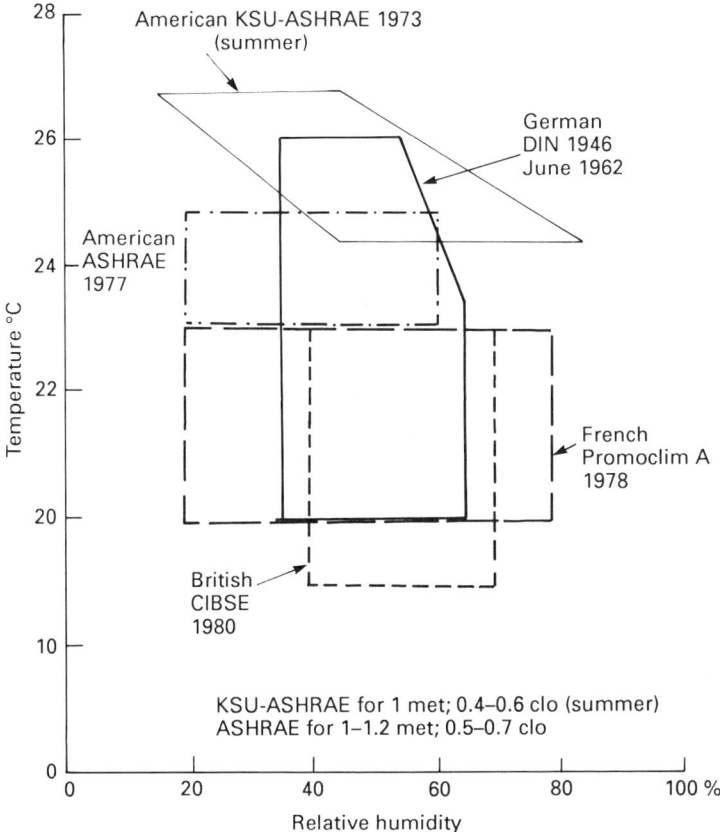

Figure 2.23 Each country chooses its own comfort envelope

2.11 Overall perception of moisture

Humidity or the effects of it on our clothing, skin, eyes, nose and throat can be perceived quite independently of other comfort factors. However, our perception is strongly influenced by temperature and activity level and there are very wide differences between people in response to any given environment. An illustration of how 32 individuals assessed the conditions inside an environmental chamber shows this. They wore standardized cotton clothing of 0.6 clo thermal insulation. They were undertaking similar sedentary activity. The room was at 30% r.h., 23°C. Their votes after three hours are shown in *Figure 2.24*. The largest number of votes were divided almost equally between 'dry' and 'normal' moisture but there were four extreme votes, two reporting 'very dry' and two 'humid'.

Spot measurements in offices show a reliable relation between ambient relative humidity and people's perception of moisture. This is illustrated in *Figure 2.25*[81].

More recent work is revealing a sensitive relationship between ambient temperature and perception of moisture. A comprehensive laboratory

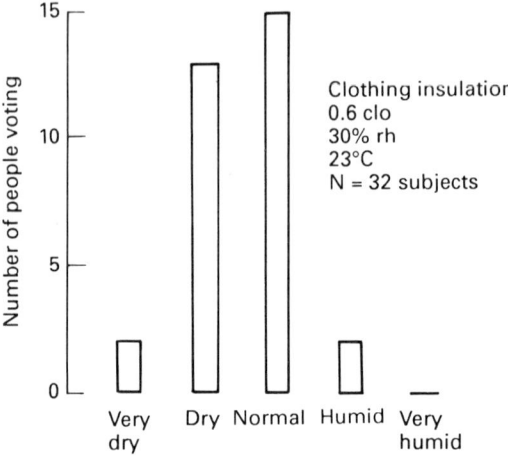

Figure 2.24 An illustration of the interpersonal differences in moisture assessment

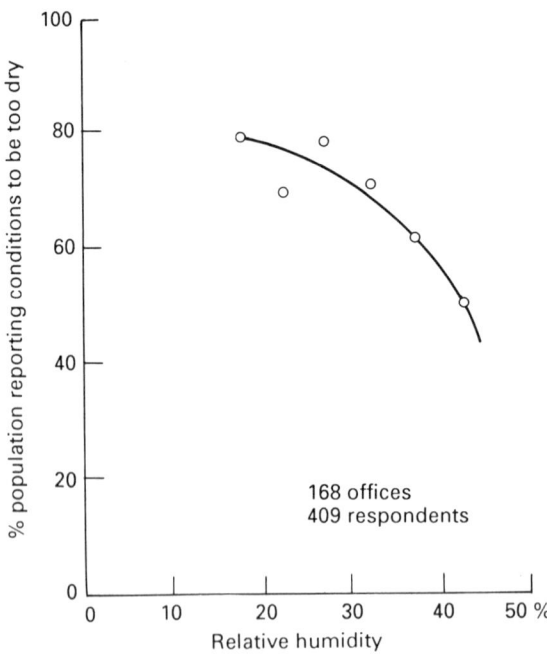

Figure 2.25 Office workers' perception of dryness with relative humidity

study used 256 subjects each once for a three-hour exposure[82]. A re-analysis of the data showed a strong relationship between moisture levels regarded as 'normal' and ambient temperature[3]. This relationship is given in *Figure 2.26* for the two relative humidity conditions tested. This suggests that comfortable moisture conditions were more strongly linked to temperature than to relative humidity or even thermal comfort. Optimum thermal comfort occurred at 25.6°C for these conditions.

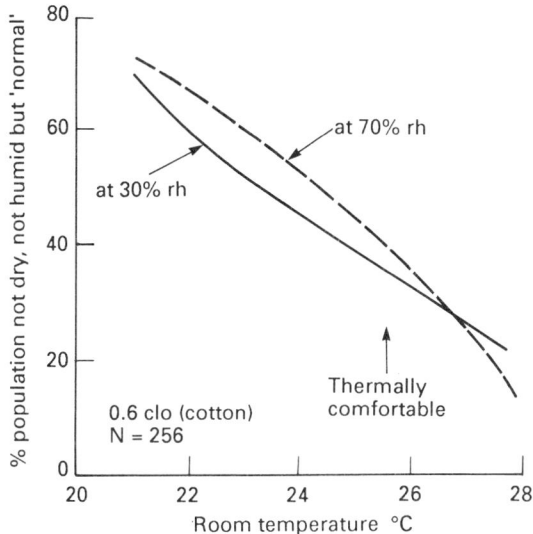

Figure 2.26 'Normal' moisture assessments change with temperature

Large-scale experiments involving four Swedish office blocks and over 600 workers pursued this line of inquiry. The internal relative humidity of the building was manipulated without informing the occupants. The occupants' views on both warmth and moisture were sought for each of four environmental conditions. The first was with full humidification, which gave around 40% r.h. in the buildings. The second was without any humidification, which resulted in 25% r.h. The other two conditions were with intermittent operation of the humidifier for one hour or two hours per day. This gave a swing in relative humidity over the working day between the two extremes of 25% and 40%. The occupants preferred their optimum temperature for thermal comfort to be 21°C, but surprisingly 86% of the measured temperatures exceeded this value by at least 1°C. This meant that most of the occupants were normally too warm. Under these conditions, at room temperatures of 23–24°C humidity increased the warmth vote and substantially increased the proportion who rated the room as 'stuffy'.

In the rooms at 21–22°C, 80% of the occupants were satisfied with the humidity level of the non-humidified rooms (25% r.h.), compared with less than 50% for those humidified (40% r.h.). Unfortunately the sample size was small.

In the rooms at 23–24°C, over 60% of the occupants found the non-humidified conditions too dry. Humidification reduced this to 30% but had the disadvantage of increasing the proportion of those who found it too humid from 30% to 60%. This meant that the total level of dissatisfaction remained constant.

The conclusion from this experiment was to maintain office temperatures at 20–22°C and ignore humidification. However, if the temperature had to be 23–24°C then intermittent humidification would please those who liked it dry and those who liked it humid for at least part of the day.

There is the possibility of a seasonal effect of moisture perception. Laboratory studies suggest that people are more conscious of dryness in winter even when the internal conditions do not change[3]. One office study reported such a seasonal effect, with 50% of the occupants reporting the office to be too dry in winter compared with 20% for the same indoor condition but in summer[83]. Another office survey found the opposite effect[84]. It therefore appears that the seasonal effect is lost in normal variations of activity and clothing amongst office staff.

2.12 Conclusions

Moisture plays a very important role in our comfort and health. Both laboratory tests and field studies show our sensitivity and dislike of both very dry and very humid conditions. In dry conditions our noses become stuffy, our lips and throats become dry, our eyes itch and become scratchy and eventually our scalp, hair and skin become dry and brittle. The time taken for these phenomena to occur varies from 15 minutes to an hour for eye dryness to a few hours for body dryness to months for an overall feeling of unpleasant dryness. In humid conditions our skin becomes damp, our clothes stick, the atmosphere becomes oppressive and sultry, and we tend to overheat. These phenomena appear more quickly than the symptoms of dryness.

These factors are mapped out in the comfort envelope shown in *Figure 2.9.*

Personal sensitivity to dryness varies widely from person to person and is strongly influenced by other factors such as temperature, clothing and activity. It also appears to be influenced by the state of health. Women are generally more sensitive than men.

Indirect effects of moisture, such as the influence on the amount of cigarette smoke generated from a cigarette, or on the probability of generating an electrostatic shock, are also important.

The research evidence suggests that low comfort temperatures of 20°C and relatively high relative humidities of 65% are the most pleasant. The modern trend to operate offices at higher temperatures appears to go against the occupants' wishes and makes it difficult to achieve comfortable moisture conditions. Humidification equipment is also treated with caution in those countries where litigation is common. Some contractors are apprehensive that any cases of humidifier fever which are caused by a dirty water circuit may be, at least in part, blamed on them.

References

1 McIntyre, D.A. 'Response to atmospheric humidity at comfortable air temperatures: a comparison of three experiments'. *Ann Occup Hyg,* **21**, 177–190, 1978
2 Andersen, I., Lundqvist, G.R. and Proctor, J. 'Human perception of humidity under four controlled conditions'. *Arch Environ Health,* **26**, 22–27, 1971
3 Andersson, L.O., Frisk, P., Lofstedt, B. and Wyon, D.P. 'Human responses to dry humidified and intermittently humidified air in large office buildings'. Swedish Building Research Report D11: 1975

4 Koch, W., Jennings, B.H. and Humphreys, C.M. 'Environmental Study II – Sensation responses to temperature and humidity under still air conditions in the comfort range'. *ASHRAE Trans,* **66**, 264–287, 1960
5 Kraemer, Sieverts and Partners. *Open Plan Offices: New Ideas, Experience and Improvements.* McGraw-Hill (UK) Ltd, London, 1977
6 Ewert, G. 'Mucus flow rate and relative humidity'. *Int Rhinology – Kyoto Congress Part I,* **4**, 25–32, 1966
7 Lake, B. and Lloyd-Hughes, J. 'Moisture studies in the domestic environment. 1. Dampness perception in laundered articles'. *J Consumer Studies Home Economics,* **4**, 87–95, 1980
8 Brundrett, G.W. 'A review of factors influencing electrostatic shocks in offices'. *J Electrostatics,* **2**, 295–315, 1977
9 Fanger, P.O. *Thermal Comfort.* Danish Technical Press, Copenhagen, 1970
10 Andersen, I., Lundqvist, G.R. and Proctor, D.F. 'Human perception of humidity under four controlled conditions'. *Arch Environ Health,* **26**, 22–27, 1971
11 Carleton, W.M. and Welch, B. 'Fluid balance in artificial environments'. NASA Report CR-114977, USAF School of Aerospace Medicine, 1971
12 Andersen, I. 'Byggematerialers afgivelse af giftige gasarter', pp. 11–12 in *Proc Symp Interclima 76.* Danish Building Research Institute, Horsholm, 1975
13 Recknagel-Sprenger. *Manuel pratique du genie climatique (Practical Manual on Climate).* PYC Edition, Paris, 1980
14 Winslow, C.E.A., Herrington, L.P. and Gagge, A.P. 'The reactions of the clothed human body to variations in atmospheric humidity'. *Am J Physiol,* **124**, 692–703, 1938
15 Miura, U. 'The effect of variations in relative humidity upon skin temperature and sense of comfort'. *Am J Hyg,* **13**, 432–459, 1931
16 Freeman, H. and Lengyel, B.A. 'The effects of high humidity on skin temperature at cool and warm conditions'. *J Nutr,* **17**, 43–52, 1939
17 Inouye, T., Hick, F.K., Telser, S.E. and Keeton, R.W. 'Effect of relative humidity on heat loss of men exposed to environments of 80, 76 and 72°F'. *Am Soc Htg Vent Engrs,* **59**, 329–346, 1953
18 Nevins, R.G., Rohles, F.H., Springer, W. and Feyerherm, A.M. 'Temperature humidity chart for thermal comfort of seated persons'. *ASHRAE Trans,* **72**, 283–291, 1966
19 Winslow, C.E.A., Herrington, L.P. and Gagge, A.P. 'Physiological reactions of the human body to various atmopsheric humidities'. *J Physiol,* **120**, 288–299, 1937
20 McK. Kerslake, D. *The Stress of Hot Environments.* Cambridge University Press, Cambridge, 1972
21 DIN 1946. *T1 Luftungsteehnische Anlagen Grundregein (Ventilation Plants. Basic Rules),* Germany, 1963
22 Raiss, W. *Heiz und Klimatechnik (Heating and Air Conditioning).* Springer-Verlag, Berlin, 1968
23 Landsberg, H.E. 'Humidity'. *Encyclopaedia Britannica,* **9**, 15th edition, 1974
24 *Rheinisch-Westfalisches Elektrizitatswerk AG,* Bau-handbuch Technischer Ausbau, RWE Essen, 1988
25 Thom, E.C. 'A new concept for cooling degree days'. *Air Conditioning, Heating and Ventilating,* **54**, 73–80, 1957
26 McArdle, B., Dunham, W., Holling, H.E., Ladell, W.S.S., Scott, J.W., Thomson, M.L. and Weiner, J.S. 'The prediction of the physiological effects of warm and hot environments'. Medical Research Council, London, RNP Report 47/391, 1947
27 Proetz, A.W. 'Humidity, a problem in air conditioning'. *Ann Otol (St Louis),* **65**, 376–384, 1956
28 Andersen, I., Luyndqvist, G.R., Jensen, P.L. and Proctor, D.F. 'Human response to 78 hour exposure to dry air'. *Arch Environ Health,* **29**, 319–324, 1974
29 Fritze, J. 'Torhedsproblemer: ovre luftveje under jetflyvning'. *Militaerlaegen,* **72**, 13–21, 1966
30 Wolff, E. *The Anatomy of the Eye and Orbit.* H.K. Lewis & Co., London, 1954
31 Duke-Elder, S. *System of Ophthalmology IV Physiology of the Eye and Vision.* Kimpton, London, 1968
32 Bell, G.H., Davidson, J.N. and Scarborough, H. *Textbook of Physiology and Biochemistry.* E. & S. Livingstone Ltd, Edinburgh, 1968
33 Holly, F.J. and Lemp, M.A. 'Tear physiology and dry eyes'. *Surv Ophthalmol,* **22**, 69–87, 1977

34 Mishima, S. and Maurice, D.M. 'The oily layer of the tear film, and evaporation from the corneal surface'. *Exp Eye Res,* **1**, 34–35, 1961

35 Mishima, S. 'Some physiological aspects of the precorneal tear film'. *Arch Ophthal Mol,* **73**, 233–241, 1965

36 Lemp, M.A., Dohlman, C.H. and Holly, F.J. 'Corneal desiccation despite normal tear volume'. *Ann Ophthalmol,* **2**, 258–284, 1970

37 Dohlman, C.H., Lemp, M.A. and English, F.P. 'Dry eye syndromes'. *Int Ophthalmol Clin,* **10**, 215–351, 1970

38 Lemp, M.A. and Hamill, J.R. 'Factors affecting tear film break up in normal eyes'. *Arch Ophthal Mol,* **89**, 103–105, 1973

39 Franck, C. 'Eye symptoms and signs in buildings with indoor climate problems'. *Acta Ophthalmol,* **64**, 306–311, 1986

40 McIntyre, D.A. 'Subjective responses to relative humidity; a second experiment'. ECRC Report N790. Electricity Council Research Centre, Capenhurst, Chester, 1975

41 Wyon, N.M. and Wyon, D.P. 'Measurement of acute response to draught in the eye'. *Acta Ophthalmol,* **65**, 385–392, 1987

42 Laviana, J.E., Rohles, F.H. and Bullock, P.E. 'Humidity comfort and contact lenses'. *ASHRAE,* **94**, Part I, Paper 3113, pp. 3–11, 1988

43 Nilsson, S.E.G. and Andersson, L. 'Contact lens wear in dry environments'. *Acta Ophthalmol,* **64**, 221–225, 1986

44 Andrasko, G. and Schoessler, J.P. 'The effect of relative humidity on the dehydration of soft lenses on the eye'. *Int Contact Lens Clinic,* **7**, 210–212, 1980

45 Hill, R.M. 'Dehydration deficits'. *Int Contact Lens Clin,* **10**, 364–365, 1983

46 Sams, W.M. 'Humidity: its relation to problems in dermatology'. *Southern Med J,* **44**, 140–147, 1951

47 Blank, I.H. 'Factors which influence the water content of the stratum corneum'. *J Invest Dermatol,* **18**(6), 433–440, 1952

48 Flesch, P. 'Chemical basis of emollient function in horny layers'. *Proc Toilet Goods Assoc,* **40**, 12–15, Dec 1963

49 Buettner, K.J.K. 'Diffusion of water vapour through small areas of human skin in normal environment'. *J Appl Physiol,* **14**, 269–275, 1959

50 Snyder, F.H. 'Glycerine and the biology of the skin'. *Drug Cosmet Industry,* **86**, 172–175, 246, 256, 1960

51 Gaul, E. and Underwood, G.B. 'Relation of dewpoint and barometric pressure to chapping of normal skin'. *J Int Dermatol,* **19**, 9–19, 1952

52 Chernosky, M.E. 'Pruritic skin disease and summer air conditioning'. *JAMA,* **1979**(13), 1005–1010, 1962

53 McIntyre, D.A. and Griffiths, I.D. 'Subjective responses to atmospheric humidity'. *Environ Res,* **9**, 66–75, 1975

54 Houghten, F.C., Teague, W.W., Miller, W.E. and Yant, W.P. 'Heat and moisture losses from the human body and their relation to air conditioning problems'. *Am Soc Htg Vent Engrs,* **35**, 245–268, 1929

55 Urquhart, A.R. 'Sorption isotherms', pp. 14–32 in *Moisture in Textiles* (ed. J.W.S. Hearle and R.H. Peters). Butterworths Publications Ltd and Textile Institute, London, 1960

56 Rees, W.H. 'The transmission of heat through textile fabrics'. *J Textile Inst,* **32**, T149–165, 1941

57 Hock, C.W., Sookne, A.M. and Harris, M. 'Thermal properties of moist fabrics'. *Am Dyestuffs Reporter,* **33**, 206–209, 1944

58 Lofstedt, B.E. 'Methods for measuring and expressing the permeability of clothing materials for wind heat and humidity'. *Biometereology,* **2**, 683–692, 1967

59 Fourt, L. and Hollies, N.R.S. *Clothing Comfort and Function.* Marcel Dekker Inc., New York, 1970

60 Seppanen, O., McNall, P.E., Munson, D.M. and Sprague, C.H. 'Thermal insulating values for typical indoor clothing ensembles'. ASHRAE Research Report 2219, RP 43, Part I, 1972

61 Fonseca, G.F. and Hoge, H.J. 'The thermal conductivity of a multilayer sample of underwear material under a variety of experimental conditions. Tech Report PR-8. Pioneering Res Div, Quartermaster R or Eng Center, Natick, Massachusetts, AD 290 744, October 1962. In *Clothing Comfort and Function,* (ed. L. Fourt and N.R.S. Hollies). Marcel Dekker, Inc., New York, 1970

62 Behmann, F.W. and Meissner, H.D. 'Die bekleidungsphysiologische Bedeutung des

Feuchteverhattens von textilen Faserstoffen'. *Melliand Textilberichte,* **40**, 1209–1214, 1959

63 Hall, J.F. and Polte, J.W. 'Effect of water content and compression on clothing insulation'. *J Appl Physiol,* **8**, 539–545, 1956

64 Flugge, C. *Microorganisms with Speical Reference to the Etiology of Infective Diseases.* Translated by W . Watson Cheyne, New Sydenham Soc., London, 1890

65 Boyce, P.R., Williams, D.W. and Brundrett, G.W. 'Lighting maintenance in an air conditioned office'. ECRC/N636. Electricity Council Research Centre, Capenhurst, 1973

66 Newsome, J.R. and Keith, C.H. Quoted in Wynder, E.L. and Hoffman, D. *Tobacco and Tobacco Smoke.* Academic Press, New York, 1967

67 Brundrett, G.W. 'Ventilation requirements in rooms occupied by smokers: a review'. ECRC/M870. Electricity Council Research Centre, Capenhurst, Chester, December 1975

68 Jensen, C.O. and Haley, D.E. 'Studies on the nicotine content of cigarette smoke'. *J Agric Res,* **51**(3), 267–276, 1935

69 Neurath, G., Ehmke, H. and Schneemann, H. 'Uber den Wassergehalt von Haupt und nobenstromrauch'. *Beitrage zur Tabak Forschung,* **3**(5), 351–357, 1966

70 Halfpenny, P.F. and Starett, P.S. 'Control of odour and irritation due to cigarette smoking aboard aircraft'. *ASHRAE,* **3**, 39–44, 1961

71 Johansson, C.R. and Ronge, H. 'Akuta irritationseffeckter av tobaksrok i rumsluft'. *Novd Hyg T,* **46**, 45–60, 1962

72 Johansson, C.R. and Ronge, H. 'Klimatinverkan pa lukt och irritationseffekt av tubaksrok. Preliminart meddelande'. *Nord Hyg T,* **47**, 33–39, 1965

73 Kerka, W.F. and Humphreys, C.M. 'Temperature and humidity effects on odour perception'. *Trans ASHRAE,* **65**, 531–552, 1956

74 Andersen, I. 'Indoor pollution due to chipboard used as a construction material'. *Atmos Environ,* **9**, 1121–1127, 1975

75 Gover, M., Reed, L.J. and Collins, S.D. 'Time distribution of common colds and its relation to corresponding weather conditions'. *Public Health Reports,* **49**, 811–824, 1934

76 Kler, J.H. 'Fewer colds in air conditioned places'. *Heating Piping and Air Conditioning,* **17**(7), 384, 1945

77 Lubart, J. 'The common cold and humidity imbalance'. *NY J Med,* **62**, 816–819, 1962

78 Sale, C.S. 'Reducing upper respiratory trace infections by allergy control and humidity control'. *Virginia Med Monthly,* **96**, 368–371, 1969

79 Gelperin, A. 'Humidification and upper respiratory infection incidence'. *Heating Piping and Air Conditioning,* **45**(3), 77–78, 1973

80 Green, G.H. 'Indoor humidity and respiratory health'. *Respiratory Technology,* **11**(3), 18–22, 1975

81 Grandjean, E. 'Die physiologische Gestalgung des Raumklimas in Aufenhaltsraumen'. *VDI-Berichte,* **106**, 29–35, 1966

82 Rasmussen, O.B. 'Man's subjective perception of air humidity', pp. 79–86, in *Proc 5th International Congress for Heating Ventilating and Air Conditioning,* Copenhagen, May 1971, Vol. I, 1971

83 Franzen, B. 'A study of the climate in nine office blocks'. Statens Institut for Byggnadsforskning Rapport 21. Stockholm, 1969

84 Nemecek, J. and Grandjean, E. 'Arbeitsphysiologische Untersuchungen der Umweltfaktoren in Grossraumburos'. *Der Arztliche Dienst,* **33**, 4–13, 1972

85 Winslow, C. E. A. The physiological influence of atmospheric humidity. *JIHVE,* **9**, 369–376, (1942)

Further reading

Andrasko, G. 'The effect of humidity on the dehydration of soft contact lenses on the eye'. *Int Contact Lens Clin,* Sept/Oct, 30–33, 1980

Andrasko, G. and Schoessler, J.P. 'The effect of humidity on the dehydration of soft contact lenses on the eye', *Int Contact Lens Clin,* **7**, 210–212, 1980

Baetjer, A.M. 'Role of environmental temperature and humidity on susceptibility to disease'. *Arch Environ Health,* **16**, 535–541, 1968

Behmann, F.W. 'The influence of climatic and textile factors on the heat loss in drying of moist clothing', pp. 273–279 in *Biometeorology.* Pergamon Press, New York, 1962

Belding, H.S., Russell, H.D., Darling, R.C. and Falk, G.E. 'Effect of moisture on clothing requirements in cold weather'. OQMG Report 37 War Department Contract 44-109-qm-445. Harvard Fatique Laboratory, Cambridge, Massachusetts, July 1945

Berglund, L.G. and Gonzalez, R.R. 'Evaporation of sweat from sedentary man in human environments'. *J Appl Physiol,* **42**, 767–772, 1977

Blank, I.H. 'Factors which influence the water content of the stratum corneum'. *J Invest Dermatol,* **18**, 433–440, 1952

Brown, S.I. and Mishima, S. 'The effect of blinking on tear concentration and corneal hydration'. *Invest Ophthalmol,* **4**, 946, October 1965

Buettner, K.J. 'The role of the skin barrier layers'. Symposium on Physical Biochemistry and Physical Aspects of Emollience. *Proc Toilet Goods Assoc,* **40**, December 1963

Buettner, K.J.K. 'Diffusion of water vapour through small areas of human skin in a normal environment'. *J Appl Physiol,* **14**, 269–275, 1959

Burch, G.E. and Winsor, T. 'Rate of insensible perspiration locally through living and through dead human skin'. *Arch Intern Med,* **74**, 437–444, 1944

Carter, J.T. and Ewell, D.G. Fluid imolance associated with hydrophilic lenses symptoms etiology and possible mechanisms. *J Am Optom Assoc,* **43**(3), 327–329, 1972

Chapple, C.C. 'The controlled physical environment for the premature and older infant', pp. 251–253 in *Aerobiology* (ed. F.R. Moulton. American Association of Advanced Science, No. 17, 1942

Dohlman, C.H., Lemp, M.A. and English, F.P. 'Dry eye syndromes'. *Int Ophthalmol Clin,* **10**, 215–351, 1970

Effron, N. and Carney, L.G. 'Effect of blinking on the level of oxygen beneath hard and soft contact lenses'. *J Am Optom Assoc,* **54**, 229–234, 1983

Franck, C. 'Eye symptoms and signs in buildings with indoor climate problems'. *Acta Ophthalmol,* **64**, 306–311, 1986

Gaul, L.E. 'Relation of dewpoint and barometric pressure to horny layer hydration'. *Proc Sci Section Toilet Goods Assoc,* **40**, 1–24, 1963

Goodman, A.B. and Wolf, A.V. 'Insensible water loss from human skin as a function of ambient vapour concentration'. *J Appl Physiol,* **26**(2), 203–207, 1969

Grice, K., Saltar, H. and Baker, H. 'The effect of ambient humidity on transepidermal water loss'. *J Invest Dermatol,* **58**(6), 343–346, 1972

Hardy, H.B., Ballou, J.W. and Wetmore, O.C. 'The prediction of equilibrium thermal comfort from physical data on fabrics'. *Textile Res J,* **23**(1), 1–10, 1953

Hertzman, A.B., Randall, W.C., Peiss, C.N. and Seckendorf, R. 'Regional rates of evaporation from the skin at various environment temperatures'. US Air Force Tech. Report AF 6680, 1951

Hertzman, A.B., Randall, W.C., Peiss, C.N. and Seckendorf, R. 'Regional rates of evaporation from the skin at various environmental temperatures'. *J Appl Physiol,* **5**, 153–161, 1953

Hollies, N.R.S. and Bogaty, H. 'Some thermal properties of fabrics. part 2. The influence of water content'. *Textile Res J,* **35**, 187–290, 1965

Iwata, S., Lemp, M.A., Holly, F.J. and Dohlman, C.H. 'Evaporation rate of water from the precorneal tear film and cornea in the rabbit'. *Invest Ophthalmol,* **8**, 613–619, 1964

Johnson, C. and Shuster, S. 'The measurement of transepidermal water loss'. *Br J Dermatol,* **79**, 575–581, 1967

Johnson, H.D. 'The response of animals to heat'. *Meteorological Monographs,* **6**(28), 112–121, 1965

Kingdom, K.H. 'Relative humidity and airborne infections. *Am Rev Respir Dis,* **81**, 504–512, 1960

Koch, W. 'Humidity sensations in the thermal comfort range'. *Architect Sci Rev,* 33–34, March 1963

Kuno, Y. *Human Perspiration.* Thomas, Illinois, USA, 1956

Kuno, Y. *Human Perspiration.* Publication 285 American Lecture Series. A monograph of American Lectures in Physiology. Charles C. Thomas, USA, 1965

Lemp, M.A. and Hamill, J.R. 'Factors affecting tear film break up in normal eyes'. *Arch Ophthalmol,* **89**, 103–105, 1973

Licht, S. (ed.). *Medical Climatology.* Eliz. Licht Publisher, New Haven, USA, 1964

McCutchan, J.W. and Taylor, C.L. 'Respiratory heat exchange with varying temperature and humidity of inspired air'. US Air Force Tech. Rep. 6023, Oct 1950

McCutchan, J.W. and Taylor, C.L. 'Respiratory heat exchange with varying temperature and humidity of inspired air'. *J Appl Physiol,* **4**, 121–135, 1951

McIntyre, D.A. *Indoor Climate.* Applied Science, London, 1980

McIntyre, D.A. 'Response to atmospheric humidity at comfortable air temperature: a comparison of three experiments'. *Ann Occup Hyg,* **21**, 177–190, 1978

Mackie, I.A. and Seal, D.V. 'The questionably dry eye'. *Br J Ophthalmol,* **65**, 2–9, 1981

Mali, J.W. 'Transport of water through the epidermis'. *J Invest Dermatol,* **27**, 451–464, 1955

March, C. 'Evaluation of the meaning and measurement of hydration in the horny layer in relation to emolliency'. *Proc Toilet Goods Assoc,* **40**, 25–27, 1963

Masnick, K.B. and Holden, B. 'A study of water content and parametric variations of hydrophilic contact lenses'. *Aust J Optom,* **55**, 481–487, 1972

Nagata, H. 'Evaporation of sweat on clothed subjects'. *Japanese J Hyg,* **17**(3), 155–161, 1962

Negus, V. 'Protection of the respiratory tract'. *Br Med J,* **II**, 723–728, 1961

Nishi, Y. and Gagge, A.P. 'Moisture permeation of clothing – a factor governing thermal equilibrium and comfort'. *ASHRAE Trans,* **76**(1), 137–145, 1970

Recknagel Sprenger. *Manuel pratique du genie climatique. Traduit de l'allemand par J.L. Cauchepin.* PYC Edition, Paris, 1979

Robinson, S., Turrell, E.S. and Gerking, S.D. 'Physiologically equivalent conditions of air temperature and humidity'. *Am J Physiol,* **143**, 21–32, 1945

de Roetth, A. 'Lacrimation in normal eyes'. *Arch Ophthalmol,* **49**, 185–189, 1953

Rolando, M. and Refojo, M.F. 'Tear evaporimeter for measuring water evaporation rate from the tear film under controlled conditions in humans'. *Exp Eye Res,* **36**, 25–33, 1983

Rubin, A. 'Thermal comfort in passive solar buildings – an annotated bibliography'. Nat. Bureau of Standards NBSIR-82-2585, 1982

Scheuplein, R.J. and Blank, I.H. 'Permeability of the skin'. *Physiol Rev,* **51**(4), 702–747, 1971

Soderstrom, G.F. and DuBois, E.F. 'The water elimination through skin and respiratory passages in health and disease'. *Arch Intern Med,* **19**, 931–957, 1917

Spruit, D. and Malten, K.E. 'The influence of the humidity of the air upon the measurement of the water permeability of the skin'. *Ann Ital Derm Clin Speriment,* **23**, 93–96, 1969

Uttley, M. 'Measurement and control of perspiration'. *J Soc Cosmet Chem,* **23**, 23–43, 1972

Chapter 3

The house dust mite allergy

3.1 Introduction

The link between the house dust mite and the common problem of asthma and rhinitis has only recently been made. Asthma is a difficulty of breathing often associated with coughing. It is most distressing if it occurs at night and interrupts sleep. Rhinitis is an inflammation of the nose which can cause similar distress. Both asthma and rhinitis are now associated with an allergic response to the house dust mites and their metabolites. These mites are more numerous in damp houses[1]. They are small in size (300 μm) and while just visible to the naked eye they are usually distinguished from dust only when they move (*Figure 3.1*).

The presence of a distinct allergen in house dust has been recognized for some time[3–5]. House dust itself is a very variable mixture of skin scales, fragments of hair and clothing fibres, food traces, mould spores, fragments of insects, and outdoor debris such as soil and grit. The allergic response could be demonstrated by scratching the skin of the patient with an extract taken from the house dust. This response could be demonstrated from the dust of the patient's own house and also from dust taken from most other

Figure 3.1 The adult house dust mite (about 300 μm long) (based on electron micrograph by G.A.H. Helson in Ingham and Ingham, 1975)[2]

houses. The potency of dust from other houses was very variable but was most powerful from houses built in regions of heavy soil or a high ground water level. The dust potency declined with the height of the house above sea level for Swiss sites (*Figure 3.2*). Early Dutch studies showed that there was a clear age relationship to this particular allergy. It was most pronounced in the younger age groups. It declined rapidly above 30 years old and hardly affected the old. Approximately 60% of the asthmatic patients referred to the Department of Allergology were sensitive to house dust (*Figure 3.3*)[6].

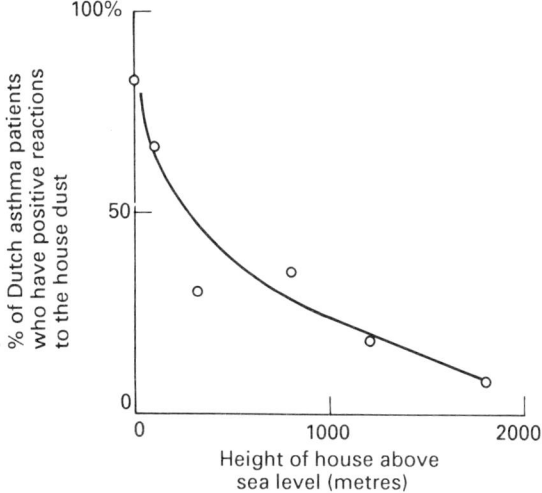

Figure 3.2 The response of Dutch asthma patients when exposed to dust from Swiss households sited at different heights

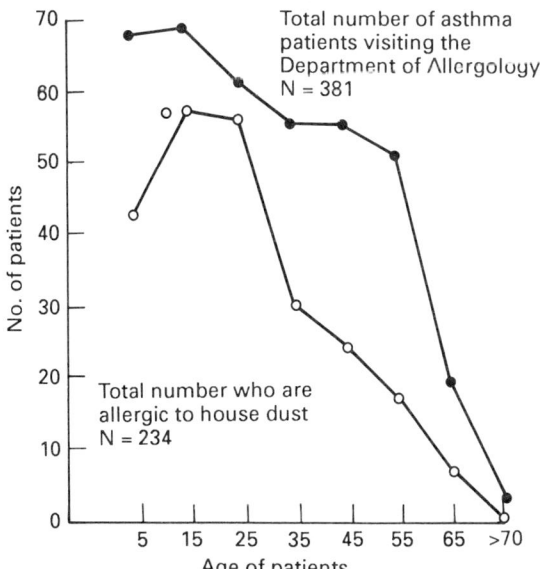

Figure 3.3 The bulk of the asthmatic patients referred to the Department of Allergology of Leiden Hospital in 1957/1958 were allergic to house dust (Voorhorst, 1962)[6]

Early American studies around the River Niagara and Lake Erie demonstrated that caddis flies (Tricoptera) were inhalant allergens to allergic patients with rhinitis and asthma[7]. However, it was not until the pioneering Dutch work that the common allergen present in all the samples of house dust was identified. This allergen was the mite and the most common type was the bed mite *Dermatophagoides pteronyssinus*[8].

Investigations in a range of Dutch houses showed a close relationship between an architect's appraisal of dampness in the house and the number of mites counted in a sample of the house dust. Dampness was based on site factors of soil type and drainage, the type of dwelling construction and whether or not it is well maintained, the heating system and its use, the arrangements for ventilation, obstruction of sunlight and exposure to wind, and living habits, including use of clothes-drying equipment and ventilation habits. The more important signs of dampness were damp spots on walls, particularly behind pictures and furniture, evidence of mildew on walls, leather and clothes, and musty smell. Samples of dust were collected from 150 ordinary houses using a vacuum cleaner and the number of mites present in a five-gram sample were counted. These results showed a six-fold increase in mites from 15 mites/5-gram sample in the dry houses to an average of 98 mites/5-gram sample from the dampest houses (*Figure 3.4*).

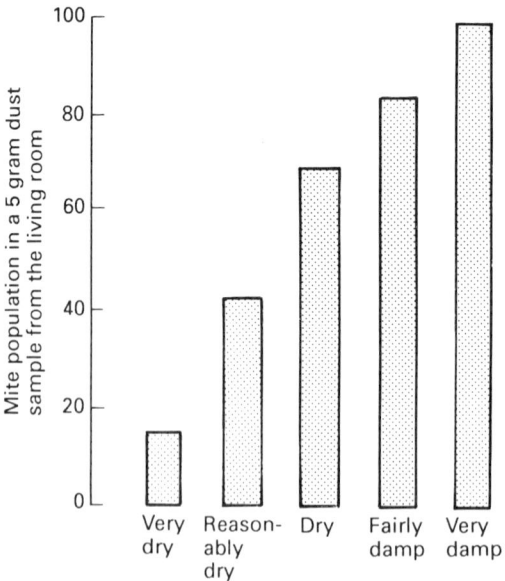

Figure 3.4 Mite population as a function of subjective assessment of house dampness (Varekamp, 1966)

Danish studies extended this work to the houses in which the allergic patients lived and compared it with a series of matching controls. Dampness was recorded objectively in terms of absolute humidity measured 200 mm above floor level. There was a clear relationship between the numbers of mites in the dust and the humidity recorded in the house. This is illustrated in *Figure 3.5*. The studies showed that there were very small numbers of mites in houses with a humidity below 6 g/kg dry air in bedrooms, or below 7 g moisture/kg dry air in living rooms. This is equivalent to a relative humidity of 60% in bedrooms at 15°C and 50% in living rooms at 20°C. Dust from the mattresses contained many more mites than dust in the living room. There was a similar increasing relationship with humidity for both the patients' houses and the control houses, but the concentrations of mites were always higher in the patients' houses. There were also more patients' houses recording the higher humidities[9].

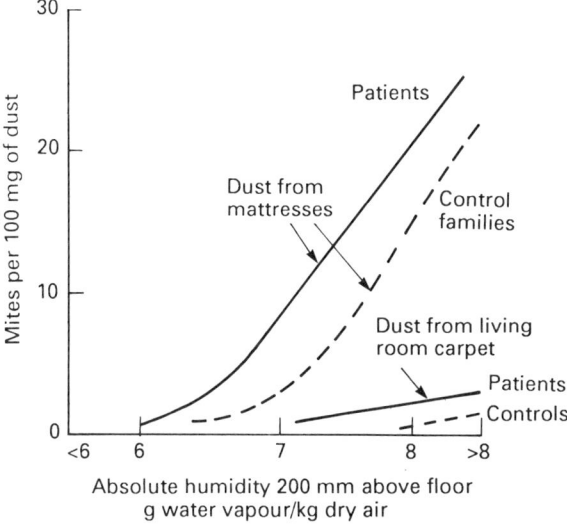

Figure 3.5 Mite concentration is linked to humidity in the room (Korsgaard, 1982)[9]

There is a seasonal pattern to the mite population. Dust samples from three houses in Leiden, The Netherlands, were collected over a year and the mite content analysed. This relationship is illustrated in *Figure 3.6*. While the concentrations of mites in each house are very different from each other, the seasonal pattern is common to all. The mite population increases to a maximum in the months of August, September and October and decreases to a minimum in the months of February, March and April[10].

We can therefore conclude that all houses contain mites in concentrations from 1 to 1000 mites per gram of dust. Damp houses will contain the largest numbers of house mites. These will be most numerous in the dust in the seams and depressions in the bedroom sleeping mattresses. Such mites and their metabolites, particularly the minute faecal pellets (20–50 μm diameter), can induce asthma and rhinitis in sensitive people. This allergic response is created by inhaling the foreign bodies and can be distressing in the short term because of the difficulty in breathing and in the longer term because of the cumulative disruption of sleep. The people most likely to be

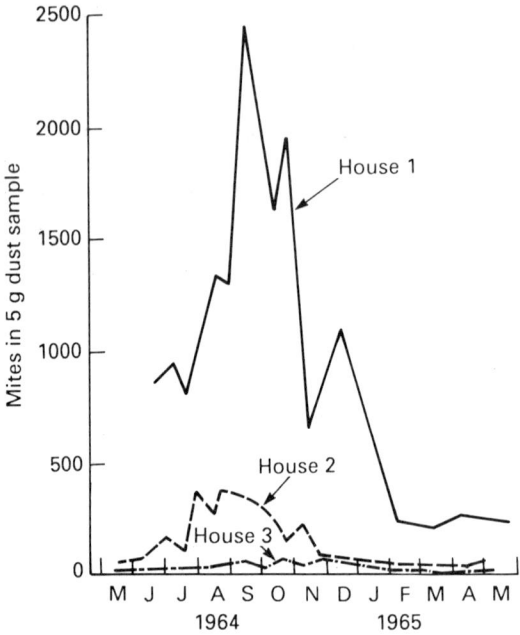

Figure 3.6 Seasonal change in the number of the house dust mites *Dermatophagoides pterenyssinus* in 5 g of dust from three houses in Leiden, The Netherlands (Spieksma, 1966)[10]

affected will be in their teens or early twenties. The probability of allergic response then rapidly declines with age.

Let us now look at the breeding cycle of the house dust mites.

3.2 The bed mite *Dermatophagoides pteronyssinus*

Mites are very small in size (250–350 μm long) with four pairs of limbs and are characterized by having a squat body which contains the head, thorax and abdomen fused together (*Figure 3.1*). They are readily identified as mites when viewed through a low-powered microscope.

Mites do not possess any special breathing apparatus such as tracheae or spiracles. The oxygen uptake and carbon dioxide release occur over the whole thin surface of the cuticle of the body. This makes the mite particularly susceptible to water loss through transpiration through this cuticle. The mites appear to detect falling humidity, and to preserve their own moisture they seek the more sheltered microclimate of a crack or crevice and then cease activity. Legs are drawn close to the body, and the underside of the body is pressed close to the surface. When large number of mites are together they also press together, often in layers two or three deep[11].

The activity and lifespan of the mite are strongly influenced by the ambient temperature. At the optimum temperature of 25°C, and at 75% r.h., ten adult mites can create almost 600 mites in six weeks. The breeding

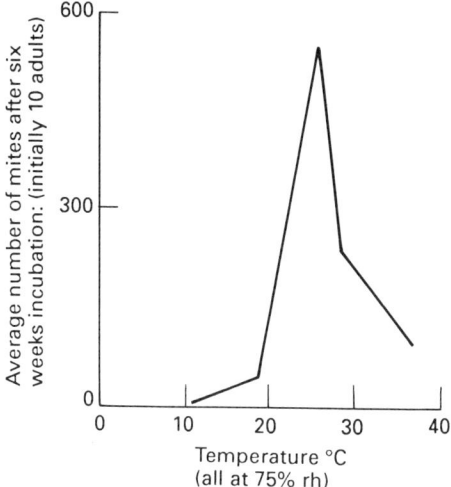

Figure 3.7 The sensitivity of the house dust mite *Dermatophagoides pteronyssinus* to temperature at the optimum relative humidity of 75% (Murton and Madden, 1977)[12]

rate is very slow below 20°C and above 40°C (*Figure 3.7*). The mite itself is tolerant to cold and can survive six weeks at 6°C and four weeks at 4°C. It can also survive for up to six hours at 51°C, 60% r.h. The relative humidity is even more critical. The optimum is 80% r.h. At lower relative humidities the mites die through desiccation. Experiments have shown that they are less sensitive to the low relative humidities when mites are present in large quantities because of their habit of clustering around each other. At higher

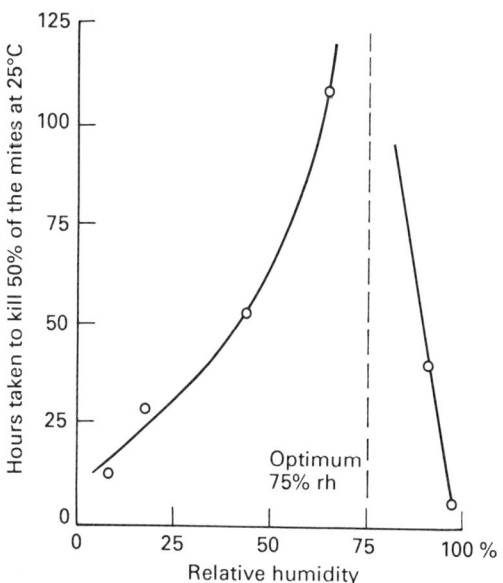

Figure 3.8 The sensitivity of the house dust mite *Dermatophagoides pteronyssinus* to relative humidity at the optimum temperature of 25°C (Murton and Madden, 1977)[12]

relative humidities, particularly above 88% r.h., there is a risk of mould which inhibits the mite development.

The favoured food of the mite from laboratory diets to date is human skin scales which have been washed in acetone and mixed with 17% dried powdered yeast.

The life cycle has been studied in the laboratory under the optimal conditions of 25°C, 80% r.h. Mating takes place two or three times during adult life. This is followed by the female laying 10–20 eggs singly at the rate of one to four per day. The eggs are large (140 μm long), elongated and initially soft, but harden after six to eight hours and adhere slightly to the substrate. After six to eight days the first larva emerges (~165 μm long). This is a small incomplete version of the adult mite but with three pairs of limbs. This larva moults after six days to form a larger version called a protonymph which, after five more days, moults to form a larger version called a tritonymph. After a further six days this moults and the full-sized adult mite emerges. This adult has a life span of just over two months[10, 11].

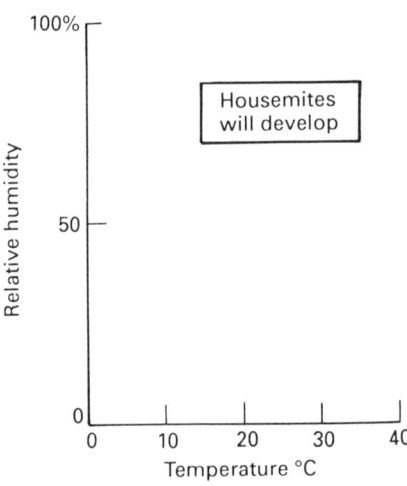

Figure 3.9 Environmental conditions for the development of the house dust mite *Dermatophagoides pteronyssinus*

These characteristics provide an understanding of the early findings on the house dust allergy. The mite needs moist conditions and prefers a temperature around 25°C. Its favourite food is based on skin scales. It is too small to be identified with the naked eye unless there is a major infestation. It can generate moisture from its own metabolism acting upon the food. While sensitive to desiccation, it has protective techniques to survive and this is particularly noticeable when mites are together in large numbers and cluster in tight groups to inhibit evaporation. They are tolerant to cold conditions, although they will not breed in them. They are well able to multiply in the mattresses of regularly used beds. The sleeping person would be expected to have a surface temperature around 33°C and to release 30 g of moisture per hour, half of which would be through respiration and the other half through the skin. Dead skin scales are shed at the rate of 0.5–1 g/week in bed[13], while the total human debris from the skin is estimated at 5 g/week[14].

3.3 Design and maintenance criteria

Three complementary steps are necessary to minimize the risk of mite infestations:

(1) *Regular, thorough cleaning.* The shortest development time of the bed mite from egg to adult mite is 20 days. The cleaning cycle has to be more frequent to break this cycle. Thorough vacuum cleaning of the mattress and the bedroom floor should be done weekly. This will remove the surface mites and remove the skin scales which constitute their food. Blanket and sheets should also be aired regularly.

(2) *Reduction in moisture generation within the house.* Any defects in the building structure which permit damp penetration from rain or from the soil should be repaired. Any moisture-releasing equipment should be properly vented. This includes clothes-drying machines and the use of flueless paraffin or gas heaters.

(3) *Prevention of mite development.* While the optimum environment is recognized as 25°C, 80% r.h., the limits at which the mites would not develop or multiply are not so clearly known. General guidelines suggest that the mites will not multiply below 15°C nor above 35°C at 75% r.h. They are unlikely to multiply below 70% r.h. and above 85% r.h.

3.4 Conclusions

Very dry conditions, as experienced in houses built in mountainous conditions in Switzerland, for example, prevent serious infestations of the house dust mite. Damp housing conditions in lowland damp climates such as Britain encourage the mite, particularly in the warmth of the bed. In these circumstances there is the risk of an allergic asthma being induced in sensitive persons, usually young people.

The most appropriate solution is to introduce a regular, thorough dust-cleaning programme which includes the mattress, reduce moisture release inside the house by venting clothes-drying machine and paraffin and gas heaters directly to the outside, and finally by maintaining a room environment below 70% r.h. by ventilation or dehumidification.

References

1 Voorhorst, R., Spieksma, F.Th.M. and Varekamp, H. *House Dust Atopy and the House Dust Mite.* Staflen's Scientific Publishing Company, Leiden, 1969
2 Ingham, P.E. and Ingham, D.M. 'Dust mites and baby care lambskins'. Wool Research Organization of New Zealand, Communication 31, April 1975
3 Kern, A. 'Dust sensitization in bronchial asthma'. *Med Clin N America,* 5, 751–756, 1921
4 Cooke, R.A. 'Studies in specific hypersensitiveness: New etiologic factors in bronchial asthma'. *J Immunol,* 7, 147–157, 1922
5 Storm van Leeuwen, W. 'Diagnosis and treatment of allergic diseases'. *Neurotherapie,* 6, 1–9, 1922
6 Voorhorst, R. *Basic Facts of Allergy.* HE Stenfert Kroese NV, Leiden, 1962
7 Parlata, S.J. 'A case of coryza and asthma due to sand flies (caddis flies)'. *J Allergy,* 3, 125–138, 1929

8 Voorhorst, R., Spieksma-Boezeman, M.I.A. and Spieksma, F.T.M. 'Is the mite the producer of the house dust allergen?' *Allergic Asthma,* **10**, 329, 1964

9 Korsgaard, J. 'Preventive measures in house dust allergy'. *Am Rev Respir Dis,* **125**, 80–84, 1982

10 Spieksma, F.Th.M. 'Some preliminary remarks on the classification, occurrence and life style of the mite *Dermatophagoides pteronyssinus*'. *Interasma,* **5**, Utrecht, 37–43, May 1966

11 Blythe, M.E. 'Some aspects of the ecological study of the house dust mites'. *Br J Dis Chest,* **70**, 3–31, 1976

12 Murton, J.J. and Madden, J.L. 'Observation on the biology, behaviour and ecology of the house dust mite *Dermatophagoides pteronyssinus*'. *J Aust Entomol Soc,* **16**, 281–287, 1977

13 Maunsell, K., Hughes, A.M. and Wraith, D.G. 'Mite asthma: cause and management'. *Practitioner,* **205**, 779–783, 1970

14 Goldschmidt, H. and Klegman, A.M. 'Quantitative estimation of keratin produced by the epidermis'. *Arch Dermatol,* **88**, 709, 1964

Further reading

Bronswijk, J.E.M.H. and Sinha, R.N. 'Pyroglyphid mites (Acari) and house dust allergy'. *J Allergy,* **26**, 31–52, 1971

Busvine, J.R. *Insects and Hygiene.* Methuen & Co., London, 1966

Charlet, L.D., Mulla, M.S. and Sanchez-Medina, M. 'Domestic acari of Columbia: abundance of the European house dust mite *Dermatophagoides pteryonssinus* in homes in Bogota. *J Med Ent,* **13**(6), 709–712, 19

Cunnington, A.M. 'House dust mites and respiratory allergy'. *Clin Allergy,* **1**, 447–449, 1971

Dixit, I.P. and Mehta, R.S. 'Prevalence of *Dermatophagoides* Sp. Bogdanov 1964 in India and its role in the causation of bronchial asthma'. *J Assoc Physicians India,* **21**, 31–37, 1973

Domrow, R. 'Seasonal variations in numbers of house dust mites in Brisbane'. *Med J Aust,* **2**, 1248–1251, 1970

Foubert, E.L. and Stier, R.A. 'Clinical investigations of mites and house dust fractions'. *Proc N Central Branch Entomol Soc Am,* **26**, 61–64, 1971

Frankland, A.W. and El-Hafny, A. 'House dust and mites as causes of inhalant allergic problems in the United Arab Republic'. *Clin Allergy,* **1**, 257–260, 1971

Gitoho, F. and Rees, H.P. 'High altitude and house dust mites'. *Br Med J,* **3**, 475–476, 1971

Haarlov, N. and Alani, M. 'House dust mites in Denmark'. *Entomol Scand,* **1**, 301–306, 1970

Halstead, D.G.H. 'Changes in the status of insect pests in storage and domestic habitats', pp. 142–153 in *Proc 1st Int Working Conference for Stored Produce Entomologists,* Savannah, USA, 1975

Harsh, G.F. 'The correlation between humidity and house dust allergy', pp. 118–123 in *Proc 1st Int Congr Allergy,* Zurich, 1951

Hughes, A.M. and Maunsell, K. 'A study of a population of house dust mite in its natural environment'. *Clin Allergy,* **3**, 127–133, 1973

Ishii, A., Takaoka, M., Ichinoe, M., Kabasawa, Y. and Ouchi, T. 'Mite fauna and fungal flora in house dust from houses of asthmatic children'. *Allergy,* **34**, 379–387, 1979

Kinnaird, C.H. 'Thermal death point of *Dermatophagoides pteronyssinus* the house dust mite'. *Acarologia,* **16**, 340–342, 1974

Koekkoek, H.H.M. and van Bronswijk, J.E.M.H. 'Temperature requirements of a house dust mite *Dermatophagoides pteronyssinus* compared with the climate in different habitats of houses'. *Entomol Exp Appl,* **15**, 438–445, 1972

Lang, J.D. and Mulla, M.S. 'Abundance of house dust mites *Dermatophagoides* sp. influenced by environmental conditions in homes in Southern California'. *Environ Entomol,* **6**, 643–646, 1977

Lecks, H.I. 'The mite and house dust allergy. A review of current knowledge and its clinical significance'. *Clin Pediatr,* **12**, 514–517, 1973

Maurya, K.R. and Jamil, Z. 'Factors affecting the distribution of house dust mites under domestic conditions in Lucknow'. *Indian J Med Res,* **72**, 284–292, 1980

Mulla, M.S., Harkrider, J.R., Galant, S.P. and Amin. L. 'Some house dust control measures and abundance of *Dermatophagoides* mites in Southern California'. *J Med Entomol,* **12**, 5–9, 1975

Murton, J.J. and Madden, J.L. 'Observation on the biology, behaviour and ecology of the house dust mite *Dermatophagoides pteronyssinus*'. *J Aust Entomol Soc,* **16**, 281–287, 1977

Nath, P., Gupta, R. and Jamil, Z. 'Mite fauna in the house dust of bronchial asthma patients'. *Indian J Med Res,* **62**, 1140–1143, 1974

Oshima, S. 'Studies on the mite fauna of the house dust of Japan and Taiwan'. *Sanitary Zoology,* **21**(1), 1–17, 19

Samsinak, K., Dusbabek, F. and Vobrazkova, E. 'Notes on the house dust mites in Czechoslovakia'. *Folia Parasitol, Prague,* **19**, 383–384, 1972

Sesay, H.R. and Dobson, R.M. 'Studies on the mite fauna of house dust in Scotland with special reference to that of bedding'. *Acarologia,* **14**, 384–385, 1972

Sharp, J.L. and Haramoto, F.H. '*Dermatophagoides pteronyssinus* (Trouessart) and other Acarina in house dust in Hawaii'. *Proc Hawaii Entomol Soc,* **20**, 583–586, 1970

Spieksma, F.Th., Zuidema, M.P. and Leupen, M.J. 'High altitude and house dust mites'. *Br Med J,* **1**, 82–84, 1971

Spieksma, F.T.M. and Spieksma-Boezeman, M.I.A. 'The mite fauna of house dust with particular reference to the house dust mite *Dermatophagoides pteronyssinus*'. *Acarologia,* **11**, 226–231, 1967

Storn van Leeuwen, W. and Butel de la Riviere, J.J. 'Hausstaub als Asthmaursaiche'. *Munch Med Wschr,* **24**, 990–997, 1929

Turos, M. 'Mites in house dust in the Stockholm area.. *Allergy,* **34**, 11–18, 1979

Varekamp, H., Spieksma, F. Th. M., Lewpen, M. J. and Lyklema, A. W. 'House dust mites in their relation to dampness of houses and the allergen content of house dust. *Paper 2.56, Interasma V Congress Proceedings,* Utrecht, The Netherlands, 1966

Wraith, D.C. and Cunnington, A.M. 'The mite and childhood asthma'. *Br Med J,* **3**, 766, 1971

Wraith, D.C., Cunnington, A.M. and Seymour, W.M. 'The role and allegenic importance of storage mites in house dust and other environments'. *Clin Allergy,* **9**, 545–561, 1979

Chapter 4

Moulds

4.1 Introduction

Mould is a form of fungus which readily grows on damp materials and creates a characteristic unpleasant smell and may eventually destroy the materials on which it grows. Fungi distinguish themselves from the higher order of plants in two ways. First they lack chlorophyll, which normal plants need to photosynthesize. This means that fungi have to depend upon organic food synthesized by other organisms. It also means that they do not need light to grow. The actual amount of food required can be minute, such as that found in fingerprints on glass. Second, they are not differentiated into roots, stems and leaves but are made up of continuous interweaving fungal threads (hyphae) usually a few micrometres in diameter. When this hyphal mat forms visible growth it is known as mould.

The air around us contains a large number of fungal spores. When such spores fall onto some appropriate nutrient where sufficient moisture is present, and providing the temperature is appropriate, they will germinate and grow. Propagation is then radial from this spot by the spread of the hyphal mat which slowly forms an expanding circle of mould[1] (*Figure 4.1*).

Spore

Germination

Growth of hypha

Network of hyphae
(the mycelium)

Visible mould

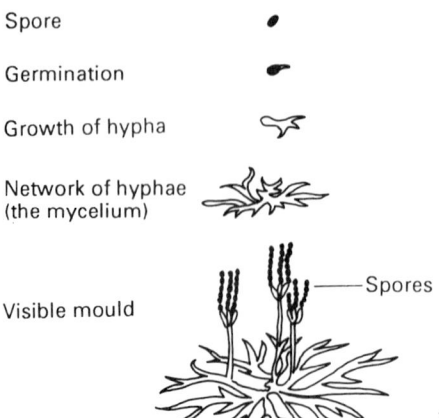

Spores

Figure 4.1 Developmental stages of a mould

This mat of fine fungal threads grows on the surface or in the substrate. Eventually the hyphae develop fruiting structures and spores themselves, which give colour to the surface of the mould patch. These spores are small, highly resistant, specialized fungal cells which take the place of seeds in the higher plants.

Propagation to other sites occurs when these spores break away and are dispersed into the air. Most spores are too small to be seen by the naked eye and usually appear as dry dust readily distributed by local air currents. Their small size and low density give them a very low falling velocity and they can therefore remain suspended in turbulent outdoor weather for several days and travel great distances. An illustration of the size and terminal velocity of common moulds is given in *Table 4.1*. A typical value for a 10-μm spore is 0.001 m/s. 'Still' air is considered to be 0.1 m/s. When these spores fall onto some appropriate nutrient which is sufficiently damp then the growth cycle restarts. Propagation can also occur from portions of the fine hyphal threads which have been severed from the original growth and blown into a new position providing nutrients.

TABLE 4.1 Size and terminal velocity for mould spores (assuming unit density and spherical shape; Mausell 1952)

Spore dia microns	Species	Terminal velocity* m/sec
3.5	*Penicillium spin.*	0.0003
2.5–4	*Aspergillus niger*	
5–10	*Cladosporium herbar*	
5–10	*Mucor racemosus*	0.0008
4–12	*Rhizopus nigricans*	
8–18	*Trichothecium roseum*	0.003
14–36	*Alternaria tenuis*	
8–35	*Macrosporium commune*	0.03

* 'still' indoor air is normally 0.1 m/s

Note: Terminal velocity is influenced by the moisture content of the spores, dry spores can fall ⅓ of the velocity of the saturated ones

4.2 Mould spores in outdoor air

The everyday outdoor biological deterioration of common plants and materials involves much digestion by mould. Since most moulds are prolific in shedding spores, outdoor air contains large numbers of spores from different fungi. In the vicinity of mildewing crops in summer the spore count can be over 10^5 per cubic metre of air, although elsewhere the concentration is more likely to be a few hundred per cubic metre of air. The actual concentration varies with the season, the time of day, the local weather, and whether the site is rural or urban. The highest values usually occur in summer[2-8]. *Cladosporium* and *Alternaria* are field fungi and show the strongest seasonal activity, peaking in summer. *Penicillium* remains relatively constant throughout the year. Within the day the spore concentrations are highest in the afternoon and lowest early in the morning[9,10].

Local weather also has an influence. Rain initially lowers the spore concentration simply by washing the spores out of the air. Prolonged rainfall, however, provides more suitable growing conditions for mould which eventually will lead to an increase in the spore concentration in the air.

Urban atmospheres contain less than half of the spore concentrations of rural sites and while both have similar types of fungal spores, the proportions of the different moulds vary. *Cladosporium, Epiccocum, Botrytis, Alternaria* and *Pullularia* dominate the rural samples, while the moulds associated more with food storage, *Penicillium* and *Aspergillus*, dominate the urban samples[2,9,11].

Local wind speed is also a factor which has a strong influence on mould spore concentration. Windy sites have lower spore concentrations. An illustrative comparison between a sheltered and a nearby exposed site over the summer months is shown in *Figure 4.2*[7].

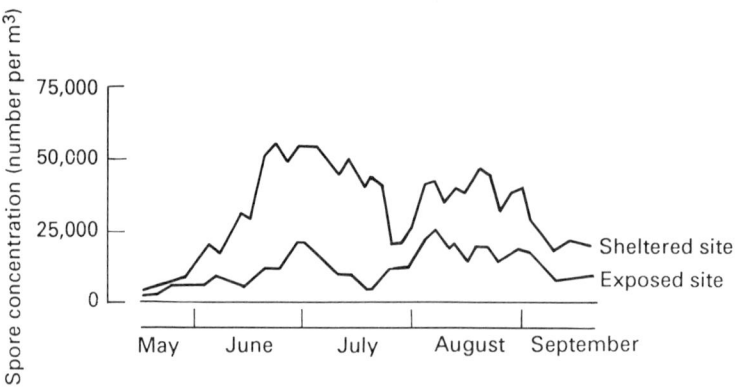

Figure 4.2 Seasonal variation in outdoor mould spore concentration

The general conclusion is that all outside air is heavily contaminated throughout the year with a rich variety of mould spores. In one study over seventy species were identified although only nine species provided 90% of the spores collected[11]. The spore concentrations are lowest in winter but rarely drop below a few hundred spores per cubic metre of air. In summer it is typically 15 000 spores/m³ but can be much higher on occasion.

Since spores are common, small in size, and can stay suspended in moving air for long periods, they will naturally enter ordinary buildings.

4.3 Mould spores indoors

Most mould spores will eventually settle onto horizontal surfaces within buildings. Household dust is therefore rich in such spores and concentrations up to 3×10^6/g of dust have been measured in homes[12]. Since mould spores are particularly resilient, they can remain dormant for several years. Some fifty-five species of mould have been identified in house dust[13]. There is agreement amongst researchers that *Aspergillus repens* is the most common household mould. It can germinate at 80% r.h. in very warm rooms. At higher humidities many more mould spores will germinate.

Indoor airborne spore concentrations are typically one-fifth of those outdoors[11, 14]. However, there is an order of magnitude difference in spore concentration between dry and damp houses and there is a distinct change in the type of mould. Damp houses had *Alternaria*, *Aspergillus*, *Ciphalosporium*, *Penicillium* and *Pullularia* moulds.

A normal house contained the same moulds as were present in the outdoor air but at a lower concentration. These moulds varied over the season. *Cladosporium* was most prevalent from June to October but *Penicillium* predominated during the rest of the year and showed little seasonal variation.

Indoor air spore concentrations also increase by over a factor of ten when dusting occurs or when building work is undertaken[15, 16].

4.4 Conditions for germination and growth

Mould spores are environmentally robust and can remain dormant and viable for many months in dry conditions. The time taken to germinate is influenced by both relative humidity and temperature. In general, the higher moisture conditions favour germination and the tolerance to moisture is widest at the optimum temperature. Growth is also influenced by the type of surface upon which the spores rest.

The germination time for *Penicillium chryogenum* on book cover materials is illustrated in *Figure 4.3*[17]. This time increases rapidly as the environmental conditions move away from the optimum of 100% r.h. at 30°C. This time can be lengthened when traces of certain chemicals are present.

The limit for mildewing is usually an arbitrary practical line[18]. Certainly, experiments on foodstuffs have shown that there is a logarithmic relation between the time taken for mould to appear and ambient relative humidity. This is illustrated in *Figure 4.4*.

Once the mould spore has germinated, the rate of growth is a similar function of temperature and humidity to that of germination[19]. Illustrative relationships are shown in *Figures 4.5, 4.6* and *4.7*. Investigations on the influence of fungicides suggest that for common materials the application of fungicides reduces the growth rate of the hyphae rather than influences the threshold moisture content at which moulds can germinate.

One critical relative humidity has been demonstrated for the mould *Aspergillus flavus*[20]. The viability of the mould spores drops from a typical value of 50% down to 10% at 75% r.h. and then becomes greater than 50% for higher relative humidities. This is attributed to a very narrow band of relative humidity just below that required for germination so that the protection of dormancy is lost but is not sufficient for growth (*Figure 4.8*).

Fungi in general have one further unusual characteristic. One of the metabolic products is water. Once growth has become established the

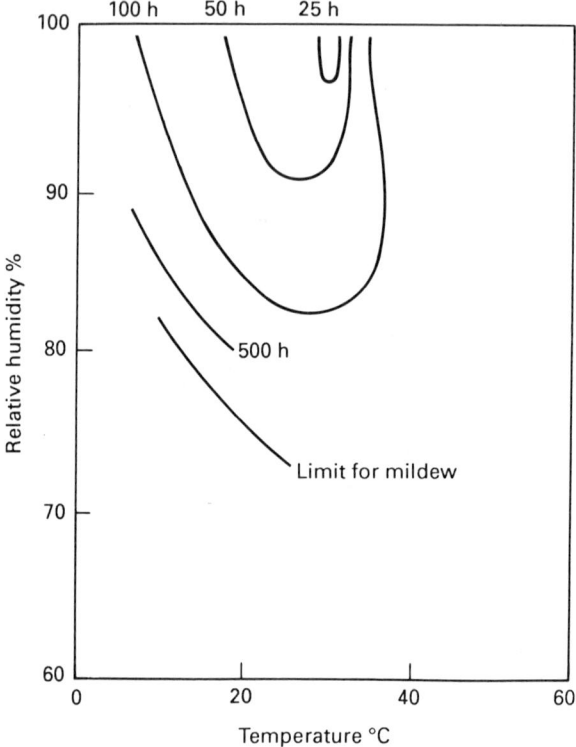

Figure 4.3 Time for mildew germination on book covers

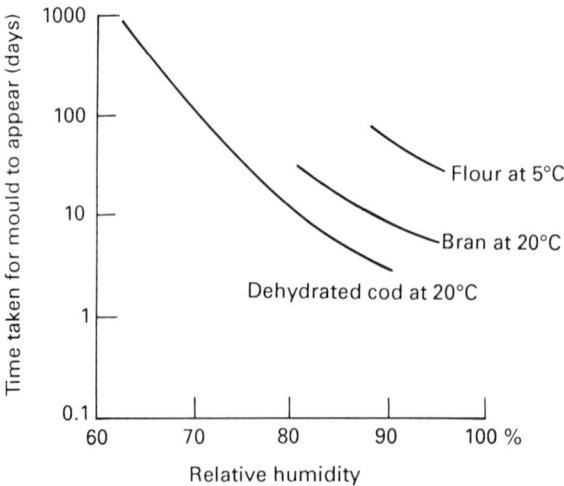

Figure 4.4 The logarithmic relationship between time taken for mould to appear on three foods and ambient relative humidity

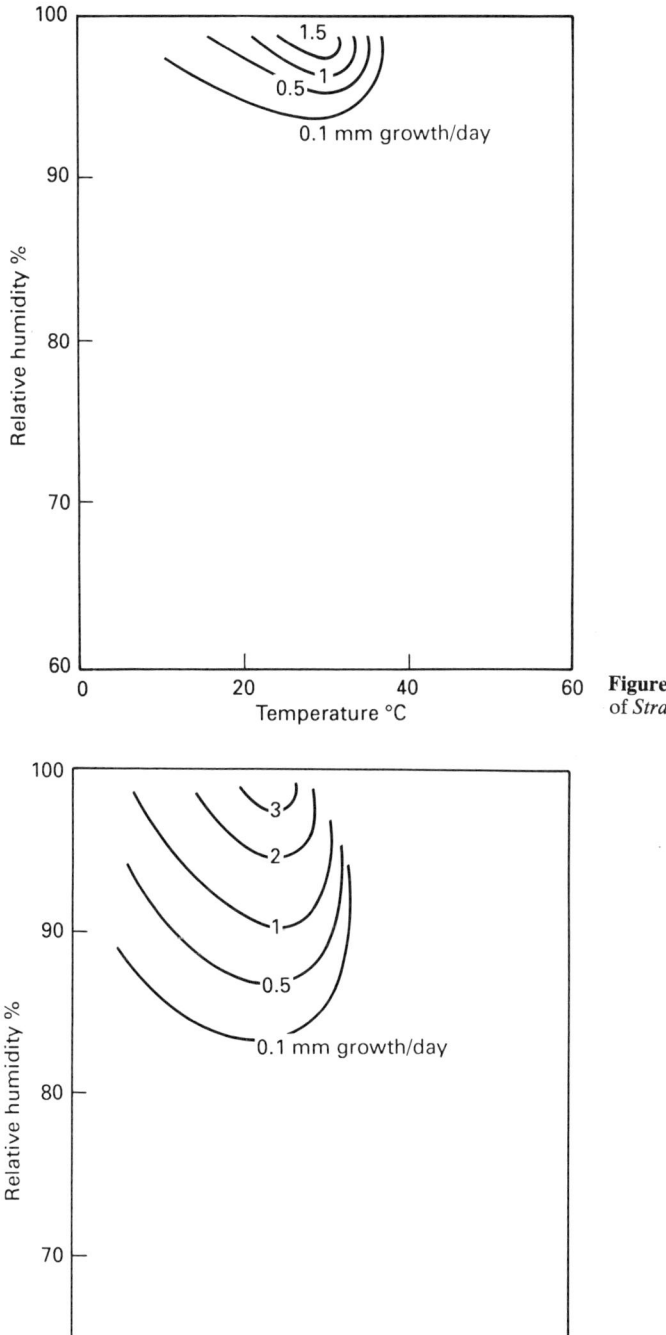

Figure 4.5 Rates of growth of *Strachybotrys atra*

Figure 4.6 Rates of growth of *Penicillium cyclopium*

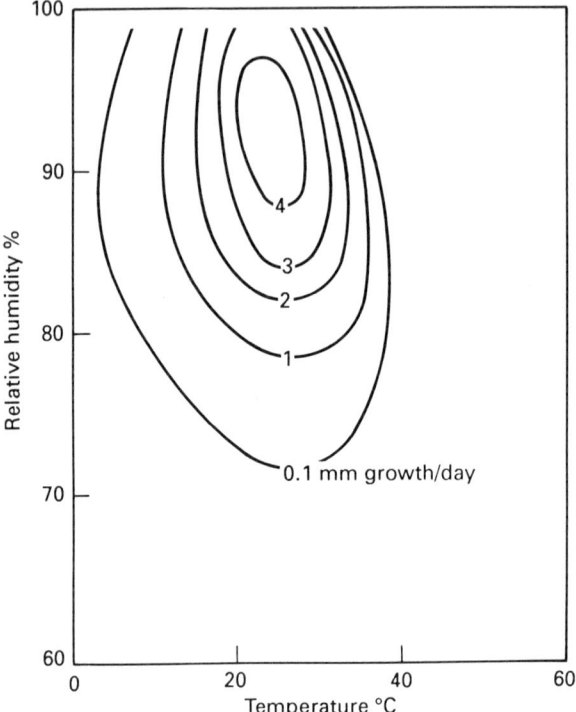

Figure 4.7 Rates of growth of *Aspergillus ruber*

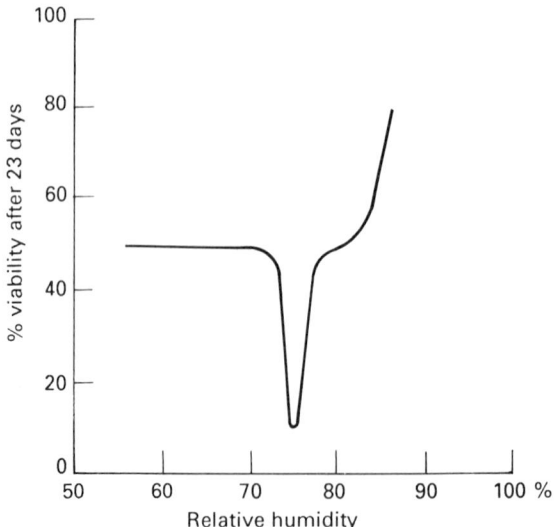

Figure 4.8 Critical relative humidity for *Aspergillus flavus*

mould structure itself can form a protective screen against desiccation. In these circumstances the mould can continue to grow even under conditions of dryness where no growth would be expected.

4.5 Sensitivity of materials to mould growth

The ease with which materials develop mould varies widely, not only with the type of material but with the type and quantity of process additives introduced during manufacture of the final product[19]. Illustrative values are given in *Figure 4.9*. Leather is one of the most widely available materials which readily grows mould, particularly *Penicillium* and *Aspergillus*, which are readily found in homes. These moulds digest the tanning material rather than the leather, although the bulk of the leather deteriorates as the tanning compounds are extracted. Goatskin, used for book covers, is the most susceptible of the leathers[17] and *Penicillium chrysogenum* could grow on it at a relative humidity of 73%.

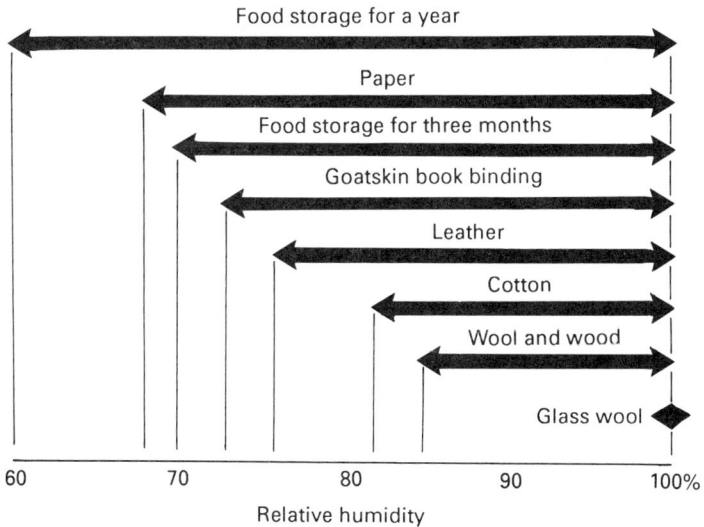

Figure 4.9 Liability to mildew for common materials

Wool is a natural protein resistant to decomposition by mould. However, the finishing process or soiling due to wear can provide nutrients for mould growth and the metabolites may react with the fabric dyes and alter the colour[21]. The lower quality wool is more susceptible to mould, and soap traces can provide nutrients. Thorough washing ensures that wool is very resistant to mould. At temperatures between 24 and 40°C there is general agreement that wool is safe from mould growth up to 85% r.h. The exact value of the upper limit of relative humidity is a function of storage time. Woollen clothing has been found to smell musty after several months at 85% r.h. even though there was no detectable growth located by microscope search[22].

Cotton is more sensitive to mould than wool and the incidence is very sensitive to temperature. Typical values of relative humidity for mould on cotton vary between 70% and 75%[23], although some authorities believe that 82% is the critical value[24]. Traces of mould have been found on cotton at 70% r.h. and 30°C although there was no growth at 5°C[25]. Mattresses smell musty after prolonged exposure to an atmosphere of 85% r.h.[19].

Mould can develop on untreated timber at 80% r.h. While in the first instance this can be harmless, the metabolic process tends to make the timber more moist. When this extra moisture exceeds 20% by weight, a value normally associated with ambient relative humidities of 85% or higher, then dry rot can occur (*Merulius lacarymans*). This particular fungus destroys timber rapidly and has the undesirable characteristic of penetrating the mortar joints of brick walls to spread to other timber.

Pure synthetic materials do not support mould growth. However, nutrients can be provided during finishing, softening and binding processes which normally occur during the production process. There is a range of tests for textile floor coverings which are designed for damp and wet areas. Individual moulds and various mixtures of moulds as recommended by the different testing bodies are used. These carpet tests showed that if one mould would grow it was extremely likely that others would in the same environment[26]. Microbial growth on plasticized PVC only occurs at the cut edges. This is attributed to the diffusion of the plasticizer on which the mould can develop. Non-diffusing plasticizers, such as phthalates, are relatively unaffected[27].

4.6 Allergies and disease

Allergens are biological substances which can initiate and provoke immunologically hypersentive reactions in some people. Allergens are proteins of relatively low molecular weight which, when suitably dispersed into the air and inhaled by susceptible persons, can induce bronchial asthma or a runny nose.

Mould spores can create allergic reactions. A simple test is to scratch the skin of a susceptible person with a needle contaminated with the mould spores. If the person is susceptible to that particular mould then the scratch will become a raised weal and will redden.

The literature over the last hundred years has linked mould spores to asthmatic attacks but there are still no agreed values on the magnitude of the problem. In a damp country such as the Netherlands[28], moulds are claimed to be responsible for 50% of the asthmatic attacks, in Germany 15%[29] and in Britain 5%[30]. Circumstantial evidence has long linked damp houses with asthmatic complaints in children. A Dutch study of 580 houses in 1969 ranked the houses in terms of dampness[31]. Two groups were selected to represent 195 dry houses and 185 damp houses with 701 and 706 children respectively. The incidence of asthmatic complaints in the children was 5.6% in the dry houses and 13.5% in the damp houses. A British appraisal of 72 asthmatic patients found that half were allergic to moulds[32] and in particular to *Penicillium*, which was the mould most frequently found in the houses. Visible moulds were noted in 26% of the homes of the asthmatic compared to 12.5% of the control group.

Clinically important concentrations of mould spores are 3000 spores/m^3 for *Cladosporium*, which is typically 5–10 μm in diameter, and 100 spores/m^3 for the larger spores of *Alternaria* (which are typically 30 μm in diameter)[33].

Special types of moulds, when found in high concentrations, can cause lung infections. The classic case is farmers' lung, which is caused by mouldy hay. When this is disturbed the mould spores are released into the air in profusion and overwhelm the natural defences of the lung. Other minority specialist lung diseases are associated with pigeon fanciers and mushroom growers.

References

1 Bravery, A.F. 'Mould and its control'. BRE Information Paper IP 11/85, 1985
2 Miquel, P. and Benoist, L. 'De l'enregistrement des poussures atmospheriques brute et organiser'. *Ann Micrographie*, **1**, 572–579, 1890
3 Feinberg, S.M. and Little, H.T. 'Studies on the relationship of micro-organisms to allergy'. *J Allergy*, **7**, 149–155, 1936
4 Durham, O.C. 'Airborne fungus spores as allergens', in *Aerobiology* (ed. F.R. Moulton). American Association of Advanced Science No. 17, 1943
5 Flensborg, E.W. and Samsoe-Jensen, T. 'Mould spore counts in outside air in Copenhagen'. *Acta Allergologica*, **1**, 104–113, 1948
6 Pasteur, V.R., Halpern, B.N., Secretain, A. and Domart, A. 'Etude de la nature et de la densité de la fiore mycologique dans l'atmosphère de Paris durant l'année'. *Acta Allergologica*, **3**, 179–197, 1950
7 Lacy, M.E. 'The summer air spora of two contrasting adjacent rural sites'. *J Gen Microbiol*, **29**, 485–501, 1962
8 Davies, R.R., Denny, M.J. and Newton, L.M. 'A comparison between the summer and autumn air spores at London and Liverpool'. *Acta Allergologica*, **18**, 131–147, 1963
9 Graham-Smith, G.S. 'The microorganisms in the air of the House of Commons'. *J Hyg*, **3**, 498–513, 1903
10 Hirst, J.M. 'Changes in atmospheric spore content: diurnal periodicity and the effects of weather'. *Trans Br Mycol Soc*, **36**, 375–393, 1953
11 Richards, M. 'A census of mould spores in the air over Britain in 1952'. *Trans Br Mycol Soc*, **39**, 431–441, 1956
12 Swaebly, M.A. and Christensen, C.M. 'Moulds in home dust, furniture stuffing and in the air within homes'. *J Allergy*, **23**, 370–374, 1952
13 Davies, R.R. 'Moulds in dusts'. *Acta Allergologica*, **13**, 124–125, 1959
14 Nilsby, I. 'Allergy to moulds in Sweden'. *Acta Allergologica*, **2**, 57–90, 1949
15 Maunsell, K. 'Airborne fungal spores before and after raising dust'. *Int Arch Allergy*, **3**, 93–102, 1952
16 Maunsell, K. Concentration of airborne spores in dwellings under normal conditions and under repair. *Int. Arch. Allergy*, **5**, 373–376, 1954
17 Groom, P. and Panisset, T. 'Studies on *Penicillium chrysogenum* in relation to temperature and relative humidity of the air'. *Ann Appl Biol*, **20**, 633–660, 1933
18 Scott, W. J. Water relations of food spoilage microorganisms. *Adv. Food Res.*, **7**, 83–127, 1957
19 Block, S.S. 'Humidity experiments for mould growth'. *Appl Microbiol*, **1**, 287–293, 1953
20 Teitell, L. 'Effects of relative humidity on viability of conidia of Aspergilli'. *Am J Bot*, **45**, 748–753, 1958
21 Burgess, R. 'Causes and prevention of mildew on wool'. *J Soc Colourists*, **50**, 138–142, 1934
22 Hartley, R.S. 'The effect of region on the liability of wool to mildew'. *Prof Textile Teachers Conference, WIRA Bull*, **9**, 29–33, 1943

56   Moulds

23 Galloway, L.D. 'The moisture requirements of mould fungi with special reference to mildew in textiles'. *J Textile Inst*, **26**, T123–T129, 1935
24 Kaswell, E.R. *Wellington Sears Handbook of Industrial Textiles*. Wellington Sears, USA, 1963
25 Illman, W.I. and Weatherburn, M.W. 'Factors affecting the development of mould on cotton fabric and related material'. *Dyestuff Reporter*, **36**, 343–344, 369–372, 1947
26 Kerner-Gang, W. and Meckel, L. 'Bestandigkeit textiler Fussbodenbelage fur Nassraume gegen'. *Mikroorganismen Melliand Textilberichte*, **53**, 1295–1298, 1972
27 Osman, J.L., Klausmeir, R.E. and Jamison, E.I. *Rate Limiting Factors in Biodeterioration of Plastic.* 66–75th Proceedings of the 2nd Institute of Biodeterioration of Materials Symposium, Vol. 2, Lunteren, Netherlands, 1971
28 van Leeuwen, W.S. 'Bronchial asthma in relation to climate'. *Proc. R Soc Med*, **17**, 19–22, 1924
29 Hansen, K. 'Uber schimmelpilz-asthma'. *Verhandl d Deutsch Gesellsch f inn Med*, **40**, 204–206, 1928
30 Hyde, H.A., Richards, M. and Williams, D.A. 'Allergy to mould spores in Britain'. *Br Med J*, **1**, 886–890, 1956
31 Leupen, M.J. Discussion: pp. 124–126 in *Indoor Climate 1979* (eds P.O. Fanger and O. Valbjorn). Danish Building Research Institute, Copenhagen, 1979
32 Burr, M.L., Mullins, J., Merrett, J.J. and Stott, N.C.H. 'Indoor moulds and asthma'. *J R Soc Health*, **108**(3), 99–101, 1988
33 Lowenstein, H., Gravesen, S. and Schwartz, B. 'Airborne allergens – identification problems and the influence of temperature, humidity and ventilation, pp. 111–123 in *Indoor Climate 1979* (eds P.O. Fanger and O. Valbjorn). Danish Building Research Institute, Copenhagen, 1979

Further reading

Albertsson, A.C. and Banhidi, Z. *Undersokning av kemiska-biologiska forandringar pa byggnadsmaterial med hygieniska och miljopaverkank – on sekvensar.* (Swedish.) *(Investigation of the chemical–biological changes in building materials with hygienic and environmental consequences.)* Department of Polymer Chemistry, Royal Institute of Technology, Stockholm, 1982
Alderson, V.G. and Mason, L.R. 'Powdery mildews as allergens'. *Californian Western Medicine*, **5**, 241–243, 1941
Ayerst, G. 'The influence of physical factors on deterioration by moulds'. *Society of Chemical Industry Monograph*, **23**, 14–20, 1966
Ayerst, G. 'The effect of moisture and temperature on growth'. *J Stored Produce Res*, **5**, 127–141, 1969
Balfour-Browne, F.L. 'Three moulds from damp walls'. *Trans Br Mycol Soc*, **35**, 273–278, 1952
Beaumont, F., Kauffman, H.F., Sluiter, H.J. and de Vries, K. 'A volumetric aerobiologic study of seasonal fungus prevalence inside and outside dwellings of asthmatic patients living in North East Netherlands'. *Ann Allergy*, **53**, 89–95, 1984
Bernton, H.S. 'Asthma due to a mould – *Aspergillus fumigatus*'. *JAMA*, **95**, 189–191, 1930
Bernton, H.S. and Thom, C. 'The role of *Cladosporium*, a common mould, in allergy'. *J Allergy*, **8**, 363–370, 1937
Betzner, M. and Lowe, E.P. 'Moulds in relation to asthma and vasomotor rhinitis'. *Mycologica*, **35**, 638–653, 1943
Beuchat, L.R. *Food and Beverage Mycology*. AVI Publishing, Connecticut, 1978
Block, S.S. 'Mould and mildew control'. Florida Engineering and Industry Experimental Station Bulletin 12. University of Florida, 1946
Block, S.S. 'Experiments in mildew prevention'. *Modern Sanitation*, **3**, 61–67, 1951
Bonner, J.T. 'A study of the temperature and humidity requirements of *Aspergillus niger*'. *Mycologia*, **80**, 728–738, 1948
Bravery, A.F. 'Mould and its control'. BRE Information Paper IP 11/85, 1985
Brijn, J. La and Kauffman, H.R. 'Fungal testing of textiles'. *Biodeterioration of Materials*, **2**, Symposium, 1972
Brown, G.T. 'Hypersensitiveness to fungi'. *J Allergy*, **7**(5), 455–470, 1937
Brundrett, G.W. and Onions, A.H.S. 'Moulds in the home'. *J Consumer Studies Home Economics*, **4**, 311–321, 1980

Buchbinder, L. 'Transmission of certain infections of respiratory origin'. *JAMA*, **118**, 718–730, 1942

Burgess, R. 'Further studies on the microbiology of wool. The enhancement of mildew by soaps and vegetable oils'. *J Textile Inst Manchester*, **20**, T333–372, 1929

Burnthall, E.V. 'The microbiological degradation of wool'. *Can Textile J*, **69**(10), 62–67, 1952

Cadham, F.T. 'Asthma due to grain rusts'. *JAMA*, **83**, 27, 1924

Chowdhury, S. 'Germination of fungal spores in relation to atmospheric humidity'. *Indian J Agric Sci*, **7**, 653–657, 1937

Christensen, C.M. 'Intramural dissemination of spores of *Hormodendrum resinae*'. *J Allergy*, **21**, 409–413, 1950

Clayton, C.N. 'The germination of fungus spores in relation to controlled humidity'. *Phytopathology*, **32**, 921–943, 1942

Cobe, H.M. 'Asthma due to a mould'. *J Allergy*, **3**, 389–391, 1932

Conant, N.F., Wagner, H.C. and Rackemann, F.M. 'Fungi found in pillows, mattresses and furniture'. *J Allergy*, **7**, 234–237, 1936

Credill, B.A. 'Report of a case of bronchial asthma due to moulds'. *J Michigan Med Soc*, **32**, 167, 1933

Creep, L.H., Tenfel, R.A. and Miller, C.S. 'Fungicidal agents in the treatment of allergy to moulds'. *J Allergy*, **29**, 258–271, 1968

Davies, R.R. 'A study of airborne *Cladosporium*'. *Trans Br Mycol Soc*, **40**(3), 409–414, 1957

Durham, O.C. 'Airborne fungus spores as allergens'. In *Aerobiology* (ed. F.R. Moulton). American Association of Advanced Science No. 17, 1943

Feinberg, S.M. and Little, H.T. 'Studies on the relationship of microorganisms to allergy'. *J Allergy*, **7**, 149–155, 1936

Fish, B.R. *Surface Contamination.* Proceedings of Conference at Gatlinburg, Tennessee, June 1964. Pergamon Press, Oxford, 1967

Flood, C.A. 'Observations on sensitivity to dust fungi in patients with asthma'. *JAMA*, **96**, 2094–2096, 1931

Galloway, L.D. 'The fungi causing mildew in cotton goods'. *J Textile Inst*, **21**, T277–286, 1930

Gilbert, R.J. and Lovelock, D.W. (eds). *Microbial Aspects of the Deterioration of Materials.* Society for Applied Bacteriology, Tech. Series No. 9. Academic Press, London, 1975

Gottlieb, D. 'The physiology of spore germination in fungi'. *Bot Rev*, **16**(5), 229–257, May 1950

Gravesen, S. 'Identification and quantification of indoor airborne micro-fungi during 12 months from 44 Danish homes'. *Acta Allergologica*, **27**, 337–354, 1972

Gravesen, S. 'Fungi as a cause of allergic disease'. *Allergy*, **34**, 135, 1979

Gregory, P.H. 'Fungus spores'. *Trans Br Mycol Soc*, **35**, 1–18, 1952

Gregory, P.H. and Monteith, J.L. (eds). *Airborne Microbes.* Society of General Microbiology, 17th Symposium, Cambridge University Press, 1967

Hamilton, E.D. 'Studies on the air spora'. *Acta Allergologica*, **13**, 143–175, 1959

Harris, L.H. 'Moulds as a cause of asthma'. *Ohio State Med J*, **34**, 158–160, 1938

Harris, L.H. 'Experimental reproduction of respiratory mould allergy'. *J Allergy*, **12**, 279–289, 1941

Hirsch, S.R. and Susman, J.A. 'One year survey of mould growth inside twelve homes'. *Ann Allergy*, **36**, 30–38, 1979

Hopkins, J.G., Benham, R.W. and Kesten, B.M. 'Asthma due to a fungus – *Alternaria*'. *JAMA*, **94**, 6–10, 1930

Hyde, H.A., 'Atmospheric pollen and spores in relation to allergy'. *Clin Allergy*, **2**, 153–179, 1972

Hyde, H.A. and Williams, D.A. 'The incidence of *Cladosporium herbarum* in the outdoor air in Cardiff 1949–1950'. *Trans Br Mycol Soc*, **36**, 260–266, 1953

Institute of Environmental Health Officers. 'Mould fungal spores – their effects on health and the control, prevention and treatment of mould growth in dwellings', Chapter 2 in *Environmental Health Professional Practice*, Vol. I. Chadwick House, London, 1985

John, A.C. and Merrett, T.G. 'The radio allergosorbent test (RAST) in nasal polyposis'. *J Laryng*, **93**, 889–898, 1979

Kalb, C.H. 'Moulds as inhalant allergens'. *Ann Allergy*, **8**, 695–698, 1950

Kanagy, J.R., Charles, A.M., Abrams, E. and Tener, R.F. 'Effects of mildew on vegetable tanned strap leather'. *J Am Leather Chemists Assoc*, **41**, 198–213, 1946

Kreuger, A.P. 'Airborne infections: a review'. *War Med*, **4**, 1–30, 1943. Naval Lab Res Unit No. 1

Lamson, R.W. and Rogers, E.L. 'Skin hypersensitivity to moulds'. *J Allergy*, **7**, 582–589, 1936

Lumpkins, E.D. and Corbit, S. 'Airborne fungi survey II Culture plate survey of the home environment'. *Ann Allergy*, **36**, 40–44, 1976

Lumpkins, E.D., Sorbit, S.L. and Tiedeman, G.M. 'Airborne fungi survey I Culture plate survey of the home environment'. *Ann Allergy*, **31**, 361–370, 1973

Madelin, M.F. (ed.). *The Fungus Spore.* Proc 18th Symposium Colston Res Soc, University of Bristol, March 28–April 1, 1966. Butterworths, London, 1946

Matthews, J.M. *The Textile Fibres.* John Wiley, New York, 1924

Merrett, T.G. and Merrett, J. 'Methods of quantifying circulating IgE'. *Clin Allergy*, **8**, 543–557, 1978

Merrett, T.G., Pantin, C.F.A., Dimond, A.H. and Merrett, J. 'Screening for IgE mediated allergy'. *Clin Allergy*, **35**, 491–501, 1980

Michaelsen, G.S. and Vesley, D. 'Dissemination of airborne microorganisms in an institutional environment', pp. 285–292 in *Surface Contamination* (ed. B.R. Fish). Pergamon Press, 1967

Miquel, P. and Benoist, L. 'De l'enregistrement des poussures atmospheriques brute et organisée'. *Ann Micrographie*, **1**, 572–579, 1890

Mislivec, P.B. and Tuite, J. 'Temperature and relative humidity requirements of species of penicillium isolated from yellow dent corn kernels'. *Mycologica*, **62**, 75–88, 1970

Mitchell Prudden, T.M.D. *Dust and its Dangers.* G.P. Putnam and Sons, New York, 1905

Morrow, M.B., Lowe, E.P. and Prince, H.E. 'Mould fungi in the etiology of respiratory allergic diseases'. *J Allergy*, **13**, 215–226, 1942

Morrow, M.B. and Lowe, E.P. 'Moulds in relation to asthma and vasomotor rhinitis'. *Mycologica*, **35**, 638–653, 1943

Mullins, J., Harvey, R. and Seaton, A. 'Sources and incidence of *Aspergillus fumigatus*'. *Clin Allergy*, **6**, 209–217, 1976

Osman, J.L., Klausmeir, R.E. and Jamison, E.I. 'Rate limiting factors in biodeterioration of plastics', in *Proc 2nd Inst Biodeterioration Symposium*, Lunteren, The Netherlands, September 1971. (eds A.H. Walters and E.H. Hueck van der Plas)

Panasenko, V.T. 'The ecology of microfungi'. *Bot Rev*, **33**(3), 189–215, 1967

Patterson, P.M. and Gay, L.N. 'The pollen content of the air and its relation to hay fever in Baltimore, Maryland during 1930'. *J Allergy*, **3**, 282, 1932

Pratt, H.N. 'Seasonal aspects of asthma and hay fever in New England with special reference to sensitivity of mould spores'. *N Engl J Med*, **219**, 782–786, 1978

Prince, H.E. 'Non pollen factors simulating seasonal respiratory allergy'. *South Med J*, **30**, 754–762, 1937

Qasem, S.A. and Christensen, C.M. 'Influence of moisture content, temperature and time on the deterioration of stored corn by fungi'. *Phytopathology*, **48**, 544–549, 1958

Rackemann, F.M., Randolph, T.G. and Guba, E.F. 'The specificity of fungus allergy'. *J Allergy*, **9**, 447–453, 1938

Richards, M. 'Atmospheric mould spores indoors and outdoors'. *J Allergy*, **25**, 429–439, 1954

Robertson, O.H. 'Airborne infection', *Science*, **93**, 213–214, 1943

Rosebury, T. *Experimental Air-borne Infection.* Microbiological Monograph: Society of American Bacteriologists. Williams & Wilkins Co., Baltimore, 1947

Rostrup, O. 'Nogle Undersogelser over Luftens Indhold af Svampekirn'. *Bot Tidsksskr*, **29**, 32–41, 1908

Saito, K. 'Untersuchungen uber die atmospherischen Pilzkeime'. *Journal of the Science College, Imperial University, Tokyo*, **18**, Article 5, 1904

Saltos, N., Saunder, N.A., Bhagwandeem, S.B. and Jarvie, B. 'Hypersensitivity pneumonitis in a mouldy house'. *Med J Aust*, **2**, 244–246, 1982

Schein, R.D. 'Comments on the moisture requirements of fungus germination'. *Phytopathology*, **54**, 1427, 1964

Schonwald, P. 'Allergenic moulds in the Pacific Northwest'. *J Allergy*, **9**, 175–179, 1938

Sherman, H. and Merksamer, D. 'Skin test reactions in mould sensitive patients in relation to presence of moulds in their homes'. *NY J Med*, **64**, 2533–2535, 1964

Snow, D. 'The germination of mould spores at controlled humidities'. *Ann Appl Biol*, **36**, 1–13, 1949

Snow, D., Crichton, M.H.G. and Wright, N.C. 'Mould deterioration of feedingstuffs in relation to humidity of storage'. *Ann Appl Biol*, **31**(2), 102–110, 1944

Solomon, W.R. 'Assessing fungus prevalence in domestic interiors'. *J Allergy Clin Immunol,* **56**, 235–242, 1975

Tomkins, R.G. 'Studies on the growth of moulds'. *Proc R Soc (B),* **105**, 375–401, 1929

Veer, A. van der. 'Mould spores in asthma and hay fever'. *J Allergy,* **8**, 277, 1937

Waldblott, G.L., Blair, K.E. and Ackley, A.B. 'An evaluation of the importance of fungi in respiratory allergy'. *J Lab Clin Med,* **26**(10), 1593–1599, 1941

Wallace, M.E., Weaver, R.H. and Scherago, M. 'A weekly mould survey of air and dust in Lexington, Kentucky'. *Ann Allergy,* **8**, 202–211, 228, 1950

Walters, A.H. and Hueck, van der Plas, E.H. *Biodeterioration of Materials,* Vol. 2. Proc 2nd Int Biodet Symp, Lunteren, The Netherlands, Spetember, 1971

Werff, van der, P.J. *Mould Fungi and Bronchial Asthma.* HE Stenfert Kroese, Leiden, 1958

Wittich, F.W. 'The nature of the various mill dust allergens'. *Lancet,* **60**, 418–422, 1940

Chapter 5

Corrosion

5.1 Mechanism

There are two basic corrosion mechanisms, namely chemical reaction and electrolytic decomposition. Chemical reactions are very temperature sensitive and at room temperatures are limited to providing a thin oxide coating, usually invisible, on the freshly prepared metal surface. This oxide layer is formed almost immediately and the metal will show little sign of further degradation in a dry atmosphere. This is termed Stage I dry corrosion and is illustrated in *Figure 5.1*. At low relative humidities there will be a monolayer of water molecules adhering to the surface[1].

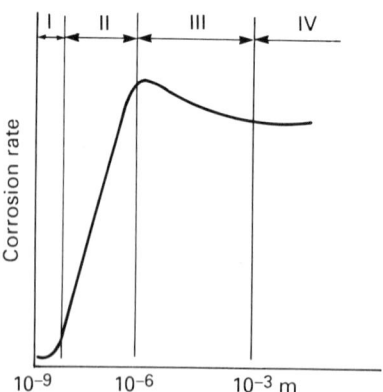

Thickness of moisture film on metal surface

I   Dry surface – very little corrosion; water molecules do not behave as water
II   Damp surface – corrosion steadily increases; molecules tend to electrolyte role
III Wet surface – thickness slightly impedes oxygen diffusion
IV Immersed surface

Figure 5.1 Illustrative zones of corrosion determined by moisture film thickness (Tomashov, 1966)[1]

As the relative humidity increases, this water layer increases exponentially in thickness *(Figure 5.2)*[2]. The thicker water film behaves progressively more like an electrolyte and establishes corrosion cells. These cells can be created by a wide variety of factors such as differential aeration, presence of different atoms in the crystal lattice, pores in the protective oxide layer or even the presence of grain boundaries. The corrosion rate increases with the thickening water film. This is termed Stage II, or moist, corrosion.

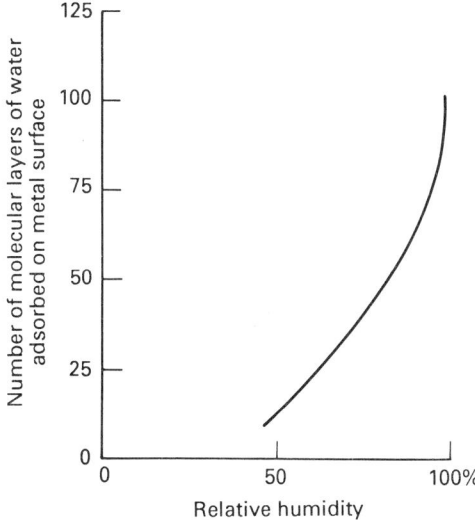

Figure 5.2 Relationship between adsorbed layers of water molecules on a clean finely polished iron surface (Tomashov and Lokotilov, 1958)[2]

At saturation, liquid condensation occurs and the droplets form a thin water film over the metal surface. The corrosion rate can fall slightly here because the thick water film can inhibit the oxygen diffusion rate to the surface. This is termed Stage III, or wet, corrosion.

Finally, the water layer is so thick that the metal can be treated as an immersed object. This is Stage IV corrosion, or total immersion. Corrosion is normally determined by the rate of oxygen reaching the metal surface.

5.2 Physical factors

The simple plain electrolyte corrosion mechanism is complicated by the physical nature of the metal surface. Lord Kelvin (1870) demonstrated that the equilibrium vapour pressure of a liquid was a function of both temperature and radius of curvature of the liquid surface. If the surface of the liquid is concave, as in a crevice in a wettable solid, then condensation could occur at much lower vapour pressures than saturation.

The thermodynamic relationship is

$$\ln \frac{P_1}{P_s} = \frac{2\sigma v}{rRT}$$

where P_1 = saturation vapour pressure above the concave liquid surface
$\quad\quad\; P_s$ = saturation varpour pressure above a flat liquid surface
$\quad\quad\; \sigma$ = surface tension of the liquid
$\quad\quad\; v$ = specific volume of the liquid
$\quad\quad\; r$ = radius of curvature of the meniscus
$\quad\quad\; R$ = gas constant
$\quad\quad\; T$ = liquid temperature (K)

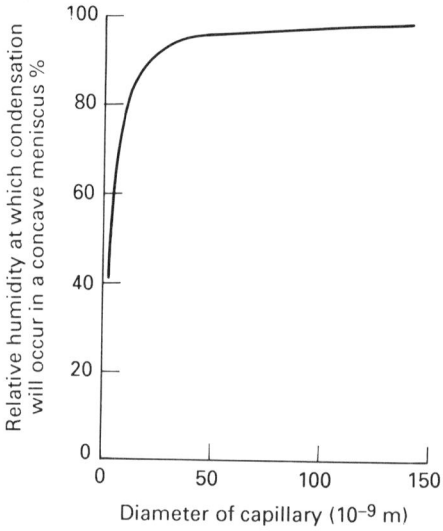

Figure 5.3 Capillary condensation can occur at relative humidities well below 100%. The critical factor is the size of the capillary (Tomashov, 1966)[1]

This relationship for water at 15°C is illustrated in *Figure 5.3* and shows how capillary condensation can occur at relative humidities well below saturation. The degree of condensation is determined partly by the size and wettable nature of the metal surface. Minutely fissured metal surfaces or metals partially protected by a porous oxide would be expected to corrode quickly because of the presence of capillary condensation at low ambient relative humidities (*Figure 5.4*).

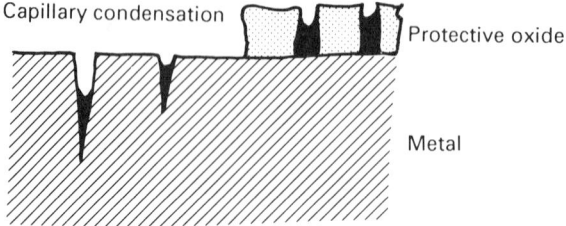

Figure 5.4 Capillary condensation can occur in fissures or micropores in the protective oxide film (Tomashov, 1966)[1]

5.3 Chemical factors

If trace chemical contaminants are present in the air then these can form a fine film of surface salt which may be hygroscopic. The presence of any soluble salt will enhance condensation of moisture because the vapour pressure above salt solutions is lower than that of pure water.

The hygroscopic characteristics of salts can be quantified in terms of the equilibrium relative humidity in a closed air space above the saturated solution. This relative humidity represents the condition where the salts dissolve in moisture absorbed from the air. Lower relative humidity values are associated with the more hygroscopic salts. The relative humidity range over which condensation will occur on different salts is illustrated in *Figure 5.5.*

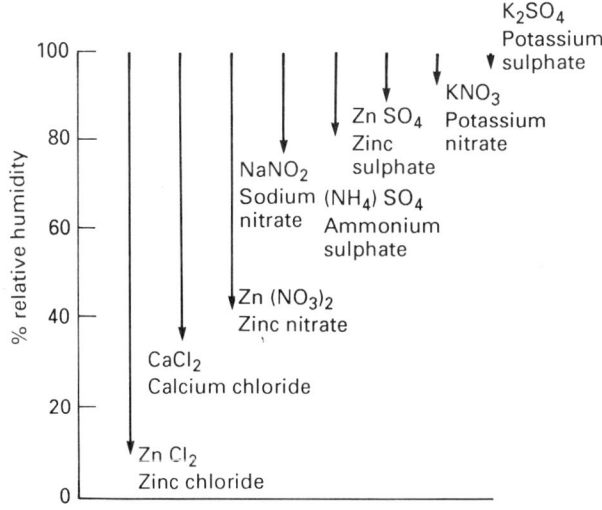

Figure 5.5 Relative humidities at which salts begin to dissolve in moisture absorbed from the air (Tomashov, 1966)[1]

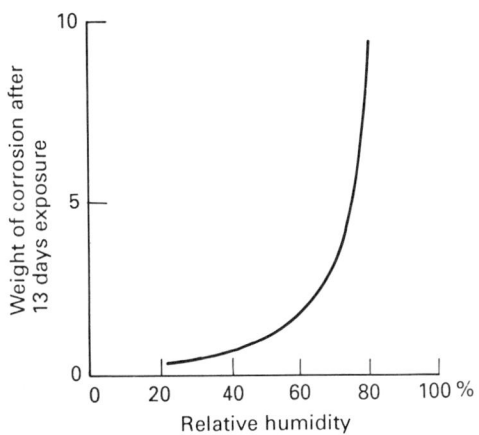

Figure 5.6 The relationship between the initial corrosion rate of steel in contact with particles of sea salt ($\sim 7\,g/cm^2$) and the ambient relative humidity (Dulz, 1950)[3]

Experiments on the corrosion rate of steel in contact with particles of sea salt indicate a critical relative humidity at 60%. The corrosion rate becomes very rapid when the relative humidity rises above 79%, the value above which the salt would be expected to dissolve and wet the metal surface (*Figure 5.6*)[3].

5.4 Critical relative humidity

The combined effect of the physical and chemical properties of a metal surface mean that there is for any given surface a critical relative humidity above which the corrosion rate becomes rapid. The actual value depends on the nature and composition of the surface. Clean iron in pure air does not corrode until the air is practically saturated. However, if the air contains a trace (0.01%) of sulphur dioxide, then the critical relative humidity is 70%. If the iron has experienced slight corrosion by exposing it to water, then the critical relative humidity is lowered to 65%. More severe prior corrosion, for example by exposing the iron to 3% sodium chloride solution, lowers the critical relative humidity to 55%.

Copper behaves similarly. The critical relative humidity is 87% after slight prior corrosion of the copper in a pure atmosphere. This value falls to 80% r.h. if the copper surface has previously been exposed to gaseous sulphur dioxide (*Figure 5.7*)[4] and to much lower values when sulphur dioxide is present (*Figure 5.8*)[5]. Traces of iodine vapour lowers the critical relative humidity to 30–40%.

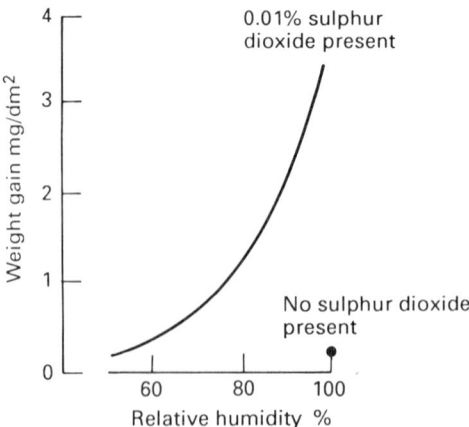

Figure 5.7 The corrosion of copper after 30 days exposure to air containing traces of sulphur dioxide and to air without sulphur dioxide (Vannerberg and Sydberger, 1970)[4]

Nickel has a critical relative humidity of 70%. Below that the metal remains bright. Higher relative humidities lead to two chemical processes. The first is the formation of a light haze, easily removed with a polishing cloth. This haze is a mixture of free sulphuric acid and nickel sulphate.

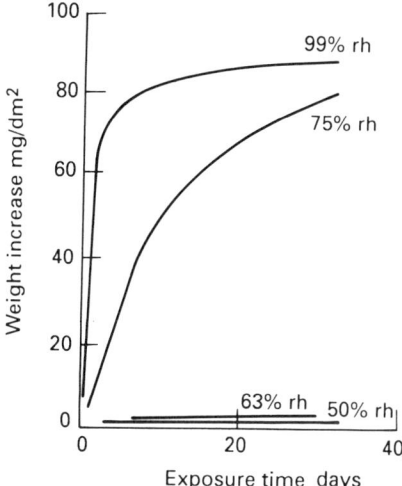

Figure 5.8 Corrosion of copper in air of various relative humidities containing 10% sulphur dioxide (Vernon, 1934)[5]

After a while this forms the solid basic nickel sulphate which can only be removed by abrasion.

Organic acids such as acetic and formic acids are released from most types of wood at high relative humidities and elevated temperatures. These acids can corrode most metals, including steel, lead, cadmium, zinc and possibly magnesium. Acetic acid, for example from wood shavings, can, in combination with the presence of carbon dioxide, attack bronze. No critical relative humidities are known. The correct solution is to avoid wood chip packaging for unprotected metallic products.

An illustration of the corrosion rates for many metals at different relative humidities is given in *Figure 5.9*[6].

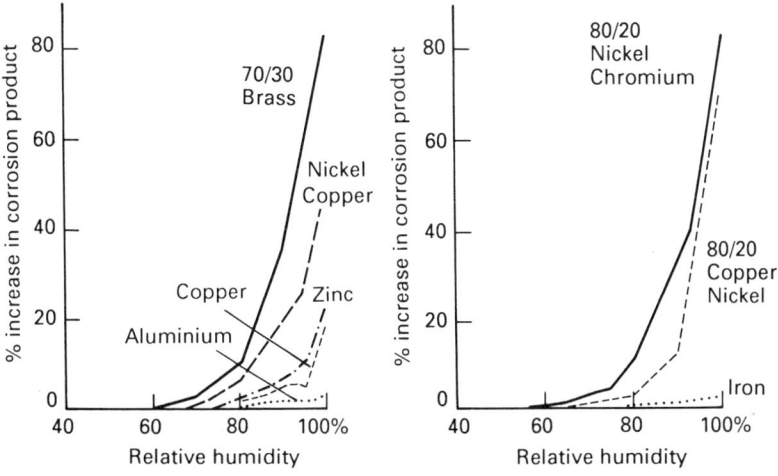

Figure 5.9 Increase in weight after exposure to successive periods of 24 hours to atmospheres of increasing humidity at 20°C (Hudson, 1929)[6]

Dust, particularly on steel, has a strong influence on the corrosion rate but less on the critical relative humidity. Vernon[5] pioneered studies on different types of dust, classifying them into three groups:

(1) Corrosion-free, e.g. an inert solid such as silica.
(2) Intrinsically corrosive – produces corrosion wherever it settles, such as ammonium sulphate.
(3) Indirectly corrisive – carbonaceous materials can adsorb acidic gases, which in turn can stimulate corrosion.

A summary of his findings is given in *Figure 5.10*[7].

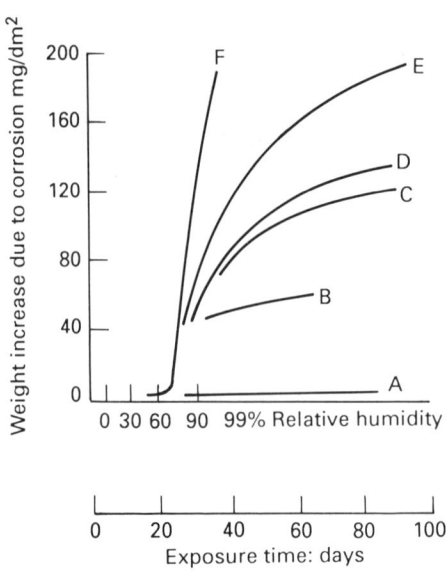

A Blank specimen or with silica particles – no sulphur dioxide
B Ammonium sulphate particles – no sulphur dioxide
C No particles – sulphur dioxide present
D Silica particles and sulphur dioxide
E Ammonium sulphate particles and sulphur dioxide
F Charcoal particles and sulphur dioxide

Figure 5.10 Critical relative humidity for iron and the subsequent corrosion rate under different types of contaminant (Vernon, 1945)[7]

5.5 Assessment of relative humidity criterion

In clean room conditions the safe criterion to avoid corrosion problems would be a maximum relative humidity of 60% for most metals. However, there will be many special industrial applications where the trace contaminants, the variability of dust composition and quantity, and the interactions between contaminants will be unpredictable in their effects on corrosion. Tomashov[1] has devised a corrosion battery which will indicate the critical relative humidity for any particular site.

The corrosion meter is made up of a sandwich of twenty or so plates of copper thinly insulated electrically from an identical number of thin iron plates. All the copper plates are wired in parallel and the iron plates similarly. The two electrical leads, one from the copper plates and the other from the iron plates, are connected to a microammeter. As the electrolytic corrosion process starts, so current will flow and be indicated on the microammeter. Research is still continuing on this meter at the Electricity Council Research Centre, Capenhurst, Chester, UK.

The construction of this meter is illustrated in *Figure 5.11* and illustrative results are shown in *Figure 5.12*.

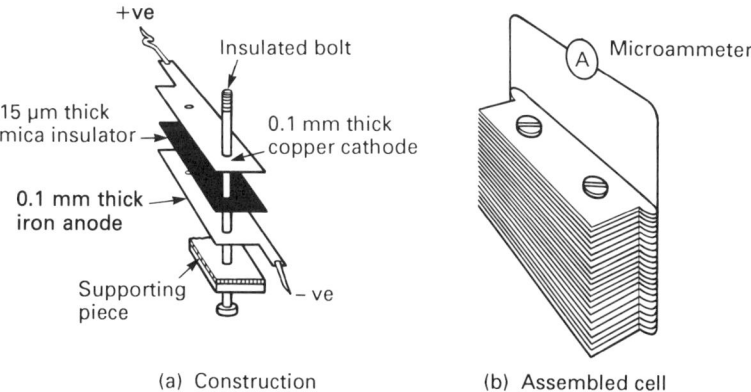

(a) Construction   (b) Assembled cell

Figure 5.11 Schematic illustration of a galvanic corrosion battery to study electrochemical corrosion processes under thin layers of adsorbed moisture (Tomashov, 1966)[1]

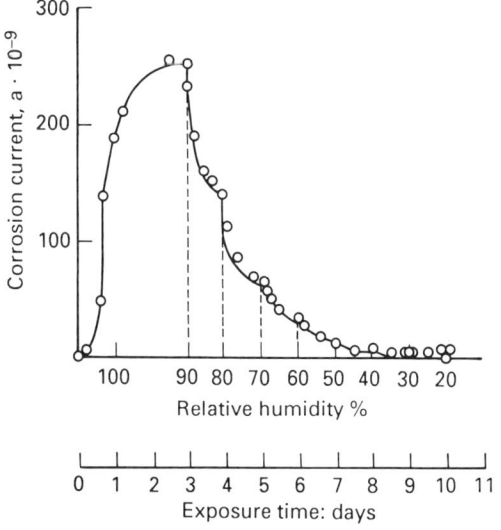

Figure 5.12 Illustration of the corrosion battery performance using copper/iron electrodes in pure air. The battery was pre-conditioned for 72 hours at a relative humidity of 100% until maximum current was reached (Tomashov, 1966)[1]

References

1 Tomashov, N.D. *Theory of Corrosion and Protection of Metals. The Science of Corrosion.* Translated by B.H. Tytell, I. Geld and H.S. Presser. McMillan Co., New York, 1966
2 Tomashov, N.D. and Lokotilov, A.A. *Zavodsk Lab,* **4**, 425, 1958
3 Dulz, S.J. *J Soc Chem Ind,* **60**, 304–315, 1950
4 Vannerburg, N.G. and Sydberger, T. *Corrosion Sci,* **10**, 43–48, 1970
5 Vernon, W.H.J. *J Chem Soc,* 1857–1868, 1934
6 Hudson, J.C. 'Atmospheric corrosion of metals: Third Report to the Atmospheric Corrosion Research Committee of the British Non Ferrous Metals Research Association'. *Trans Faraday Soc,* **25**, 177–252, 1929
7 Vernon, W.H.J. *J Sci Instr,* **22**, 228–237, 1945

Further reading

Evans, U.R. *The Corrosion and Oxidation of Metals.* Edward Arnold, London, 1960
Leidheiser, H. *The Corrosion of Copper, Tin and their Alloys.* John Wiley, New York, 1971
Tiller, A.K. 'The factors responsible for the atmospheric corrosion of metals'. Report NPK 63/16/151/0238. Electricity Council Research Centre Contract 82/8063. December 1983
Walton, J.R., Johnson, J.B. and Wood, G.C. 'Atmospheric corrosion initiation by sulphur dioxide and particulate matter'. *Br Corrosion J,* **17**(2), 59–69, 1982
Wormwell, F. *Corrosion of Metals Research, 1924–1968.* National Physical Laboratory, HMSO, 1973

Chapter 6

The influence of moisture on electrical properties

6.1 Introduction

Moisture has two kinds of effects on the electrical resistance of bodies. The first and most clearly recognized is that on the bulk resistivity of absorbent materials such as paper and textiles. The higher moisture content associated with the higher ambient relative humidities leads to easier ionic movement through the moisture electrolyte and hence to a lower electrical resistance across the material. This phenomenon is particularly important in the printing and weaving industries because of electrostatic effects associated with high-speed machinery. It is also important for comfort in offices that electrostatic shocks are avoided. The safety implications of even minute electrostatic discharges are important in explosives manufacturing and in areas specially enriched in oxygen, such as hospital intensive care units or operating theatres. There are also important cost implications of accidental discharges in the vicinity of microelectronic circuits. These special applications will be dealt with in Chapter 7.

The second moisture effect is on the surface resistance of electrical insulators. The bulk resistivity of electrical insulants is very high and effectively prohibits the conduction of electricity. However, the external surface is sensitive to both contamination and to surface moisture. This can lead to a breakdown in the electrical insulation.

Illustrative values for materials which are considered electrically conductive are less than $10^7 \Omega$ m; for those which are considered electrically insulating they are greater than $10^{10} \Omega$ m; and for antistatic materials they are between these two values[1].

The principles of the two phenomena will now be discussed.

6.2 Absorbent materials

Moisture is the most important factor in determining the electrical resistance of absorbent materials. A cotton thread can change its electrical resistance by a factor of 10^{10} over a range of ambient relative humidities from 5% to 80%. This is illustrated in *Figure 6.1*, together with data from

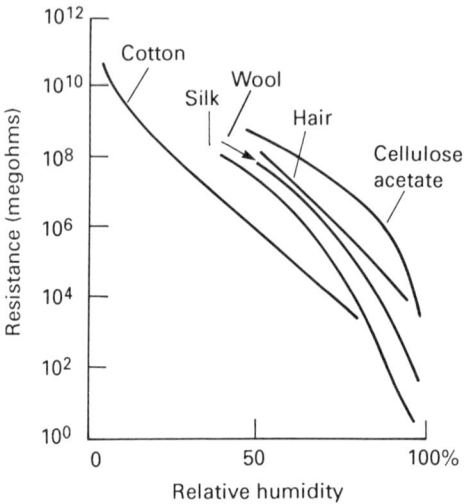

Figure 6.1 Resistance of 12.7 mm textile threads

other absorbent fibres. All show a similar change with a change in relative humidity but the absolute values vary significantly. There is no simple relationship between actual moisture content and resistance because the moisture is bound up in different ways within the different types of fibre. At a given relative humidity, for example, cotton typically holds half the moisture content of wool, and yet the electrical resistance of cotton is very much lower than that of wool[2].

There is lack of agreement amongst researchers on the reason for these electrical characteristics. The most promising theory is electrolytic conduction through the fibre. The difference between wool and cotton could be attributed to the moisture first adsorbed being firmly bound to the hydrophilic groups in the side-chains of the protein molecules of wool rather than being more loosely attached as in the cellulose structure of cotton.

When these fibres are made up into textile materials, it is difficult to use the conventional electrical definitions because they rely on knowledge of the cross-sectional area of the material. The more convenient unit is mass per unit length. The mass specific resistance is defined as the resistance in ohms between the ends of a specimen 10 mm long and of mass 1 g. Experimental measurements are illustrated in *Figure 6.2*. The results vary with the purity of the material and the particular experimental technique but the relative change in resistivity with change in relative humidity is relatively unchanged[3].

In addition to its electrically conducting properties, water is a polar molecule, which means that it can be polarized by an electric field. The degree to which it can be polarized is termed the dielectric constant. This is unity for a vacuum but increases with polarity. For most materials the value of this constant falls with increasing frequency as the larger molecules are unable to follow the rapid changes and hence no longer contribute to the dielectric constant. The dielectric constant for pure liquid water at room temperature is 81 for frequencies up to 600 MHz. Since dry

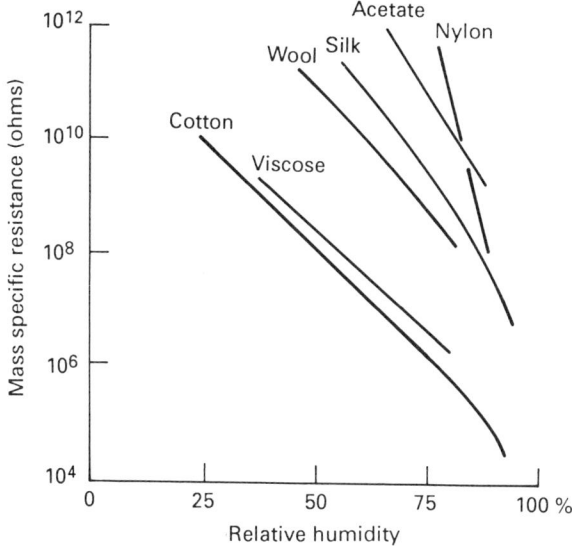

Figure 6.2 The mass specific resistance of textiles varies with relative humidity

fabrics have a low dielectric constant around 3, the addition of water enhances it. Measurements on cotton, wool and nylon are illustrated in *Figure 6.3*[3]. The reason why the effect is not more pronounced, particularly for wool, is that the water in the fibre is not all available as liquid water.

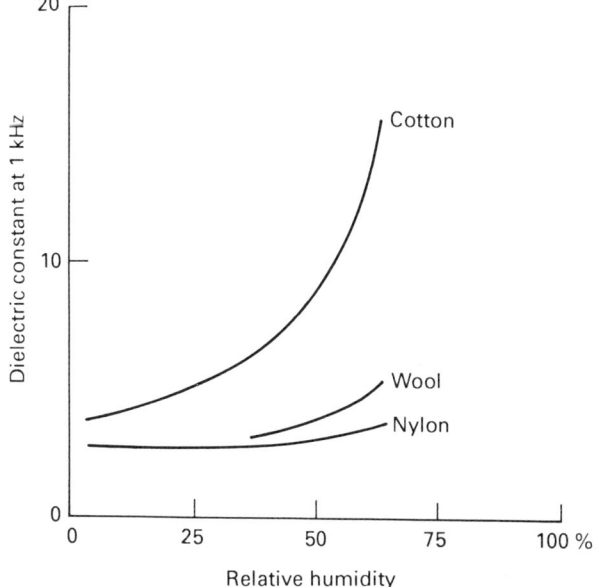

Figure 6.3 The dielectric constant of fabrics changes with relative humidity

6.3 Electrical insulation

Electrical insulators have a very high volumetric resistivity. Their effectiveness therefore depends to a large extent on the leakage current through any contaminant on the surface. Moisture is the most common contaminant. Dirt exaggerates the influence.

Molecules of water are constantly condensing upon and evaporating from every free surface. The concentration of molecules on the surface at any one time is strongly linked to the relative humidity of the air. Electrolytic conduction occurs whenever a conducting path arises[4-7]. Experiments on scrupulously cleaned flat glass showed that it required a relative humidity of 50% before the whole glass surface was covered with a single layer of water molecules[8]. More layers of water molecules grew linearly with increasing relative humidity up to 70%. The rate of growth of the molecular layers increased very rapidly above 90% r.h. This effect is illustrated in *Figure 6.4*.

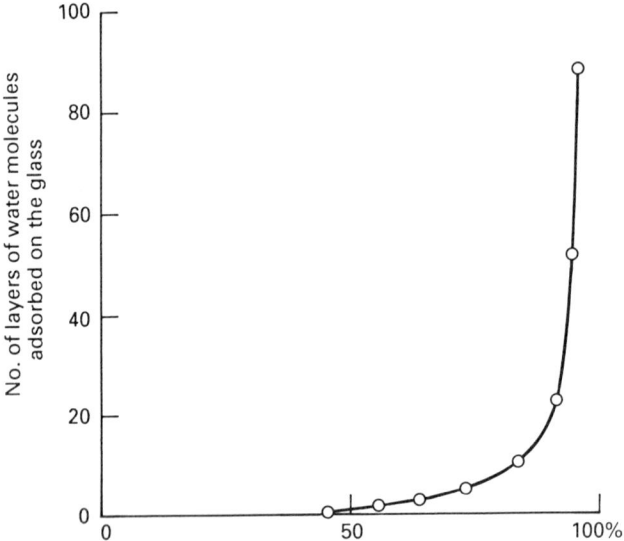

Figure 6.4 The number of layers of water molecules increases at the higher relative humidities·

The biggest single difference between insulation materials is in the form of the moisture film. The surface resistivity of clean insulating materials is illustrated in *Figure 6.5*. Some surfaces are water-repellent and any moisture collects in droplets. These materials have a very high surface resistivity and are relatively unaffected by ambient relative humidity. High molecular weight paraffin typifies this family. Other surfaces build up their moisture film in a more continuous way. Such materials have a clearly defined surface resistivity which progressively reduces with increasing relative humidity. Plate glass would typify such material. Many materials when clean and freshly prepared behave in the first water-repellent mode. After exposure to high temperature or ultraviolet irradiation, the nature of

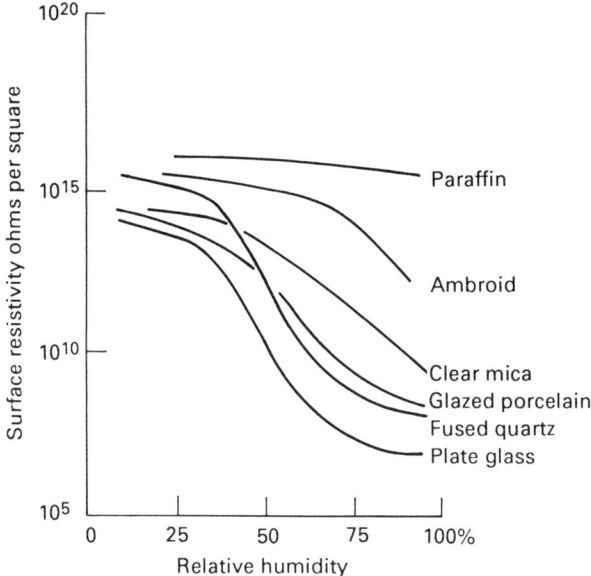

Figure 6.5 The influence of relative humidity on surface resistivity

the surface can change and it approaches the second kind of wettable surface.

If electrical equipment has to operate in very high relative humidities, then water-repellent treatment of the surface will prolong the usefulness of the wettable surface type of insulators[9]. Methyl and ethyl chlorosilanes coated over the insulator surface work well under all circumstances. The organosilicon reagents work but need careful cleaning of the glass and steatite ceramic surfaces by chromic acid before treatment[10].

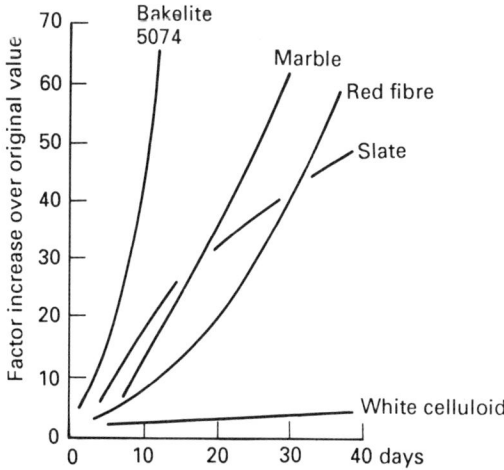

Figure 6.6 Increase in volume resistivity caused by loss of moisture in going from an ambient relative humidity of 90% to one of 25% at 25°C

Fungal growth will develop in warm humid climates. Such growth will not normally damage the components of any electrical circuit by chemical attack. However, mould is organic and damp. Once developed it can survive much lower relative humidities because water is one of its waste products of metabolism. Profuse growth of mould will lead to electrical short-circuiting.

The volumetric resistivity of materials such as slate and marble have also been shown to change with relative humidity. This is attributed to the absorption of water vapour within the material. However, such changes take days to happen (*Figure 6.6*)[11].

The breakdown of electrical insulation is also becoming a new problem at low voltages. This is due to the rapid progress in microminiaturization which is leading to very compact printed circuit boards and a widespread use of such devices. Experiments were made to weather such boards in both seaside and urban sites. Results for some of the smaller circuit spacings which were more affected than the wider ones are shown in *Figure 6.7*. The urban sites were more affected than the seaside ones and this was attributed to metallic dust. However, at 81% r.h. the breakdown voltage for both was lower than expected. In practice both the pollution and the higher relative humidities had an effect and lowered the insulation resistance. The effect was particularly pronounced above 60% r.h.

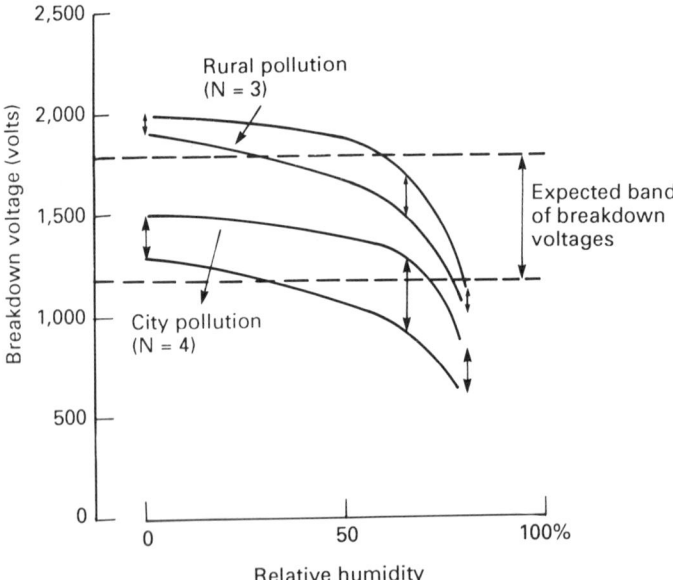

Figure 6.7 Influence of relative humidity and pollution on breakdown voltage on printed circuit boards

6.4 Conclusions

Moisture has a profound effect on the electrical resistance of non-metallic materials and the critical factor is relative humidity, not absolute humidity.

This is attributed in part to the growing thickness of the adsorbed surface moisture with increasing ambient relative humidity. It is also influenced by the absorbed moisture within the material. Changes in relative humidity from 40% to 100% can lead to a reduction in electrical resistance of 10^8 times for a fibrous thread.

This moisture film is very important. If it is not significant, as in low ambient relative humidities, then the material is an electrical insulator. This is ideal for electrical circuits but means that electrostatic charges cannot leak away and can therefore build up to a high value. This can lead to painful electrostatic shocks and to explosions in hazardous atmospheres.

If the moisture is significant, as in high ambient relative humidities, then the surfaces become much more electrically conducting. This means that there may be some leakage current in electrical circuits which may encourage malfunction or unsatisfactory operation. It also means that electrostatically generated charges leak away quickly and that spark generation is therefore unlikely.

In practice, electrostatic shocks can be eliminated by maintaining an ambient relative humidity above 50%. In hazardous areas of explosive gases or in oxygen-enriched atmospheres, ignition can occur from small sparks which would not normally be noticed by an individual. The relative humidity should be above 65% in such areas. These factors are discussed in depth in Chapter 7.

Electrical insulation is best preserved by keeping the ambient relative humidity below 60% but it is particularly important to keep the surface of the insulator clean and to prevent hygroscopic contaminants settling on the surface.

The dielectric constant of fabrics also increases with increasing ambient relative humidity. The constant for water is much higher than that for dry fabrics (81 compared with 3).

References

1 Jowett, C.E. *Electrostatics in the Electronics Environment*. Macmillan Press, London, 1976
2 Williams, R.R. and Murphy, E.J. 'The predominating influence of moisture and electrolytic materials upon textiles as insulators'. *Bell System Tech*, **2**(8), 225–242, 1929
3 Hearle, J.W.S. 'Moisture and electrical properties', pp. 123–140 in *Moisture in Textiles* (eds J.W.S. Hearle and R.H. Peters). Textile Institute/Butterworths Scientific Publications, London, 1960
4 Cohn, E.M. and Guest, P.G. 'Influence of humidity upon the resistivity of solid dielectrics and upon the dissipation of static electricity'. US Bureau of Mines Information Circular 4492, June 1944
5 Dryden, J.S. and Wilson, P.J. 'The influence of moisture on insulating materials'. *Aust J Appl Sci*, **1**, 79–112, 1950
6 Kawasaka, K., Kanoa, K. and Sekita, Y. 'Liquid like layers in the adsorbed film of water on glass'. *J Phys Soc Japan*, **13**, 222–223, 1958
7 McIllhagger, D.S. and Salthouse, E.C. 'Insulator surface conduction'. *Proc IEE*, **112**(7), 1468–1472, 1965
8 McHaffie, J.R. and Lenher, S. 'The adsorption of water from the gas phase on plane surfaces of glass and platinum'. *J Chem Soc*, **127**, 1559–1568, 1925
9 Leutritz, J. and Herman, D.B. 'The effect of high humidity and fungi on the insulation resistance of plastics'. *Bull Am Soc Textile Materials*, **138**, 25–30, 1946

10 Meakins, R.J., Mulley, J.W. and Churchward, V.R. 'Some experiments on the application of organo silicon compounds to glass and ceramic'. *Aust J Appl Sci*, **1**, 133–139, 1950
11 Curtis, H.L. 'Insulating properties of solid dielectrics'. *Bull US Bureau Standards*, **11**, 359–420, 1915

Further reading

Baker, W.P. *Electrical Insulation Measurements*. Newnes, London, 1965
Fellman, K.H., Pfeiffer, W. and Schau, P. 'Withstand voltages of small insulating paths and the effect of natural environmental conditions'. *Elektrotechnische Zeitschrift Archiv*, **3**(4), 117–120, 1981. (Health and Safety Executive Translation HSE 9693, 1981)
Gabel, M. and Schon, G. 'Electrostatic charging of paper – effect of paper velocity and surface conductivity'. *Das Papier*, **24**(7), 373–380, 1970
Gregory, P.H. *The Microbiology of the Atmosphere*. Leonard Hill Ltd, London, 1961
Hartshorn, L., Megson, N.J.L. and Rushton, E. 'Plastics and electrical insulation'. *J Inst Elec Engrs*, **83**(502), 474–496, 1938
Johnson, F.W. 'The surface resistivity of adsorbed moisture films on glazed porcelain'. *Phil Mag*, **18**, 63–80, 1934
Johnson, F.W. 'The adsorbed moisture films on the surface of glazed porcelain'. *Phil Mag*, **24**, 797–807, 1937
Kuznetsov, A. 'On the problem of surface electrical conductivity'. *Zh Fiz Khim*, **27**, 657–666, 1953
Lee, J.A. and Lowry, H.H. 'Effect of moisture on electrical properties of insulating water resins and bitumens'. *Ind Eng Chem*, **19**(2), 302–306, 1927
Ray, C.T. 'Influence of static electricity on paper printability: bibliography (1904–1973)'. Committee Assignment Report CAR No. 62, 1975. Tech Assoc of the Pulp and Paper Industry, Atlanta, USA
Weir, C.E. 'Influence of temperature and moisture on the electrical properties of leather'. *J Res Nat Bureau Standards*, **48**(5), 349–359, May 1952
Yaker, W.A. and Morgan, S.O. 'Surface leakage of pyrex glass'. *J Phys Chem*, **35**, 2026–2042, 1931

Chapter 7

Electrostatic discharges

7.1 Introduction

There are four reasons why electrostatic discharges are of growing importance:

(1) *Open-plan air-conditioned offices.* The concept of large open spaces may give the occupants the unwitting opportunity to charge themselves up to a high voltage and create painful discharges when they next touch a metallic object, such as the handles of filing cabinets[1-3].

(2) *Computer room pick-up.* The growth of highly equipped and interlinked computers in computer rooms means that unnoticed accidental discharges may introduce spurious data into the information-processing procedure and make nonsense of the data.

(3) *Manufacturing microelectronics.* Accidental electronic discharges can easily spoil finished microcircuits[4,5].

(4) *Safety.* The growing recognition of the risk in generating sparks in hazardous areas is leading to strict enforcement of antistatic precautions. This applies to conventional fuel-processing and explosions-handling industries, and also to areas where special gases such as oxygen may be present, as in hospitals[6].

Four factors influence the electrostatic discharge. These are charge generation, electrical leakage resistance, capacitance and method of discharge. In turn, the magnitude of the discharge affects the degree of nuisance and hazard. Let us examine these factors in turn.

7.2 Charge generation mechanism

Charge generation is created by the separation of dissimilar materials which have previously been in contact. During contact there is some charge transfer between the materials. As the two surfaces separate, the electrical capacitance between them falls inversely as the separation distance. Since the electrical charge is constant and

Charge Q = capacitance C × voltage V

a reduction in capacitance will produce a rise in the voltage of the surface with increasing separation distance. As a walking person lifts one foot from the floor, the increased voltage of the shoe from this effect raises the charge on the total capacitance of his whole body with a partial leak to earth through the shoe still in contact with the ground.

When the foot is lowered again it will have lost some of its charge to the main capacitance of the body and a little through earth leakage. The foot reaches the carpet at a much depleted voltage and recharges from the carpet. Thus at each step the falling and rising local capacitance between foot and carpet pumps charge into the main capacitance between body and earth, giving a ratcheting increase in personal voltage, and the higher the footsteps the greater the voltage increments. This voltage increases asymptotically with the distance walked up to some maximum value. Charging currents are estimated to be smaller than $10^{-8}\,\text{A}^7$. In steady-state conditions the maximum personal voltage will be a direct function of the leakage resistance to earth. This can be illustrated as follows:

Maximum personal voltage (volts) =
Charging current (amps) × leakage resistance (ohms)

For a given charging current the maximum personal voltage will be directly related to the leakage resistance to earth. Assuming a maximum charging current of $10^{-8}\,\text{A}$, the personal voltage will be 10 V for leakage resistance of $10^9\,\Omega$. This is illustrated in *Figure 7.1*. However, for a 1000-fold increase in leakage resistance to $10^{12}\,\Omega$, the maximum personal voltage increases also 1000-fold to 10 000 V. This leakage resistance to earth is the sum of the shoe resistance and the carpet resistance. The carpet resistance depends not only upon the environmental conditions around it,

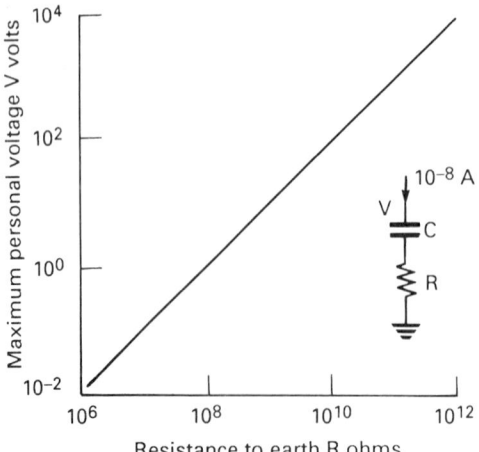

Figure 7.1 For a given charging current the maximum voltage is a function of the resistance to earth

such as moisture content, but also upon the nature of the substrate. In practice it is a complex combination of horizontal and vertical resistance along and through the carpet structure.

The electrical resistance of the shoe component of this circuit varies widely with the material and thickness of the sole of the shoe. Spot-checks on a pair of British leather-soled shoes showed the resistance to be $10^8 \Omega$, and this was unaffected by the wearing of hose[8]. A spot-check on a pair of rubber-soled plimsolls gave an electrical resistance of $5 \times 10^{11} \Omega$[9]. The electrical resistance from hand to conducting floor was measured for people wearing a range of different kinds of shoes. New dry leather soles could have resistances as high as $10^9 \Omega$, while for damp, thin soles this went down to $10^3 \Omega$. Rubber-soled shoes gave resistances of $5 \times 10^8 \Omega$, but this could go down to 10^7–10^4 with conducting rubber. PVC-soled shoes had a resistance of $4 \times 10^9 \Omega$. Microcellular rubber-soled shoes had an electrical resistance greater than $10^{13} \Omega$. Measurements of the electrical resistance of the footwear of 132 people working in hospital operating theatres showed that rubber soles predominated (81 people) and that the rest were leather-soled (51 people). The rubber soles had much higher electrical resistances than the leather ones. Seventy per cent of the rubber-soled shoes had a resistance greater than $10^9 \Omega$. Only 18% of the leather-soled shoes had a resistance greater than $5 \times 10^7 \Omega$[10]. Dry socks, particularly thick woollen ones, have a high resistance value, but this declines quickly due to perspiration which moistens the sole and lowers the electrical resistance (*Figure 7.2*)[11].

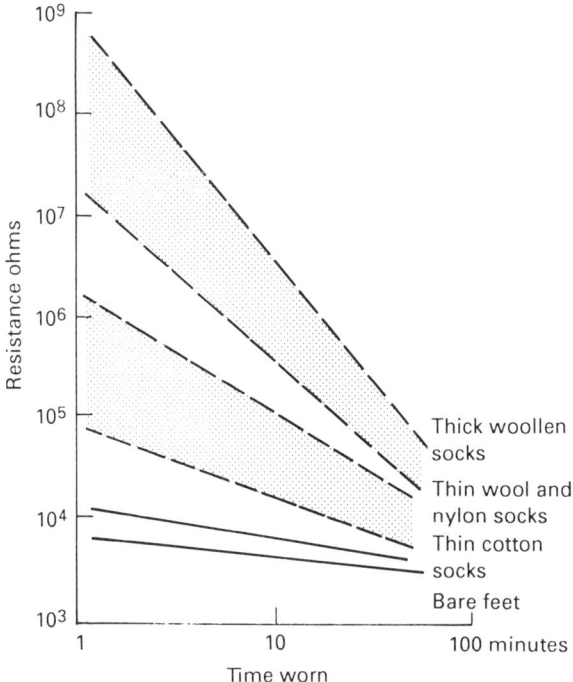

Figure 7.2 The electrical resistance of socks quickly decreases when they are worn

These results are illustrated in *Figure 7.3* and show that, with the exception of the highly insulating microcellular rubber-soled shoes, the electrical resistance of the shoes alone does not seem sufficient for the small charging current to generate a voltage sufficiently high to create an electrostatic shock in people. In general, the critical resistance in the earth path necessary to create electrostatic discharges must therefore be in the carpet or floor covering rather than in the shoes.

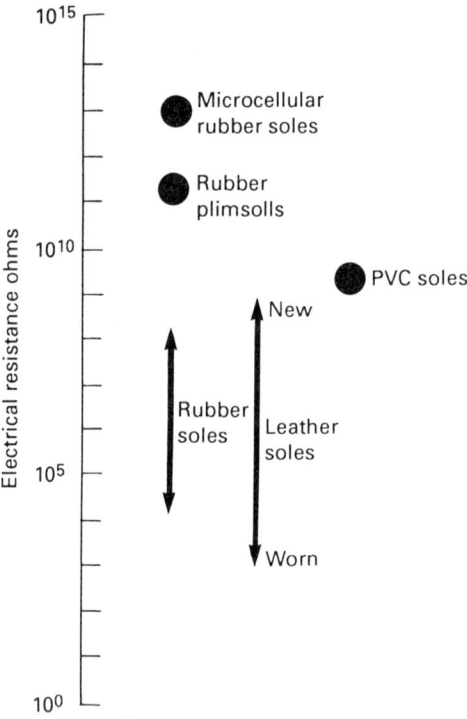

Figure 7.3 The electrical resistance of shoes varies with the sole material

Once a person stops walking he has stopped generating the charging current. The charge which has built up on the individual then leaks away. The time taken for the electrostatic charge to fall to 37% of its initial value is related to the personal capacitance and the leakage resistance:

Time constant (seconds) =
leakage resistance R (ohms) $\times$ capacitance C (farads)

This relationship is illustrated in *Figure 7.4*.

There are few measurements of the actual capacitance of a standing person. An early field survey in the USA on 22 adults recorded a range from 95 to 398 pF with an average value of 204 pF. One important factor was the thickness of the sole. Thinner soles were associated with higher capacitance[10]. Later USA studies reported values between 100 and 200 pF and noted that the value was influenced by the type and thickness of carpet

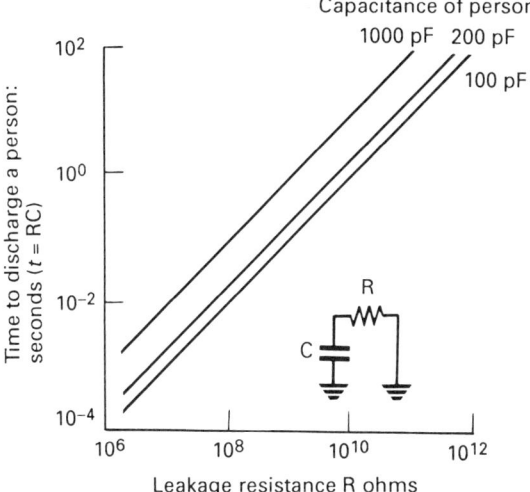

Figure 7.4 The discharge time of a person is a function of capacitance and leakage resistance to earth

on which the person was standing[12]. Early British measurements gave values between 215 and 328 pF[6,13,14] which were higher than the American results.

Research has shown that three major factors determine the electrical capacitance of a person to earth:

(1) *Thickness of the shoe sole.* Laboratory measurements show this to be the most important single factor and its influence is illustrated in *Figure 7.5*[10].

(2) *Area of sole in contact with the floor.* Men tend to have larger feet and also a larger part of their shoes is normally in contact with the ground. This

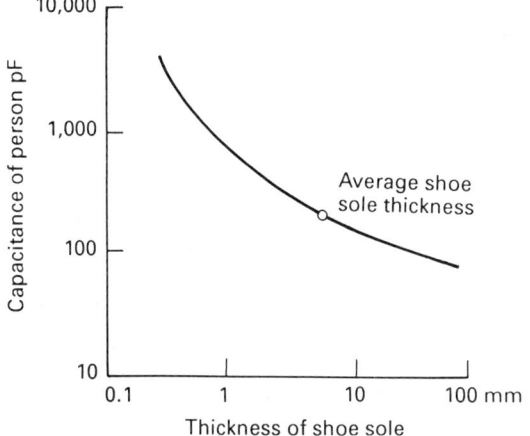

Figure 7.5 Typical curve showing the variation of a person's capacitance with thickness of shoe sole

results in the electrical capacitance of a man being higher than that of a woman.

(3) *Material of the sole.* Leather-soled shoes provide the highest capacitance. Measurements of office workers showed a range of 140–1100 pF with an average of 419 pF. Spot-checks on one pair of worn leather shoes showed a value of 470 pF, while an identical but brand-new pair provided a capacitance of 180 pF. The capacitance of men wearing rubber-soled shoes varied from 90 to 330 pF, averaging 204 pF. The capacitance for women wearing rubber-soled shoes varied from 51 to 120 pF, averaging 87 pF[15].

An illustration of the values found by researchers is given in *Figure 7.6*. The range can vary from 50 to 1100 pF, with typical values being 200 pF for men and 100 pF for women.

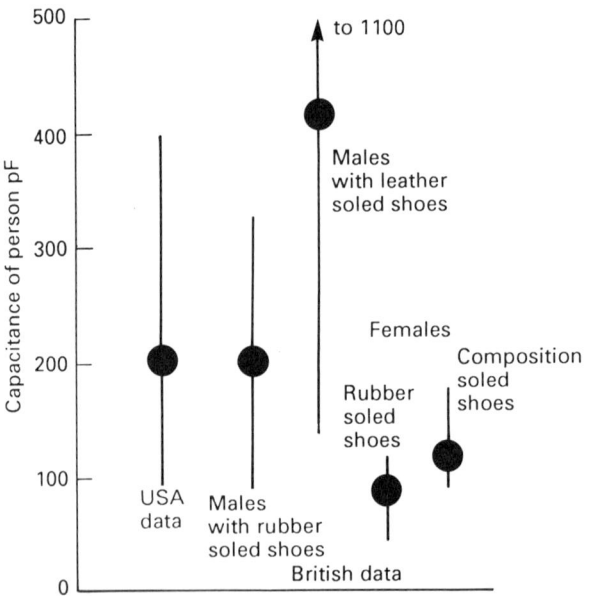

Figure 7.6 The electrical capacitance of a person is determined by the kind and thickness of the soles of the shoes worn

These values can be used in conjunction with *Figure 7.3* to assess how quickly the electrostatic charge can leak away. If the personal capacitance to earth is 200 pF and the total leakage resistance is $10^{12}\,\Omega$, then the time of discharge would be relatively long at 200 s. A lower leakage resistance of $10^{6}\,\Omega$ would result in a discharge time of only 200 μs, which is practically instantaneous for these purposes. A person would discharge immediately on pausing. One USA test for antistatic materials is to measure if a 5000-V charge could decay to less than 500 V in less than half a second[3].

The crucial factors in charging up a person electrostatically and retaining that charge for a short time when becoming stationary both depend upon the leakage resistance to earth. If this value is below $10^{10}\,\Omega$ then the personal maximum voltage would be approximately $100\,\mathrm{V}$ and the discharge time $1\,\mathrm{s}$. Since shoes are not normally so high in electrical resistance, the critical factor becomes the electrical resistance of the floor covering.

Moisture content is the most important factor in determining the electrical resistance of textile materials. The moisture content is closely related to the ambient relative humidity. This moisture relationship, termed a sorption isotherm, is illustrated in *Figure 7.7* for a range of textile

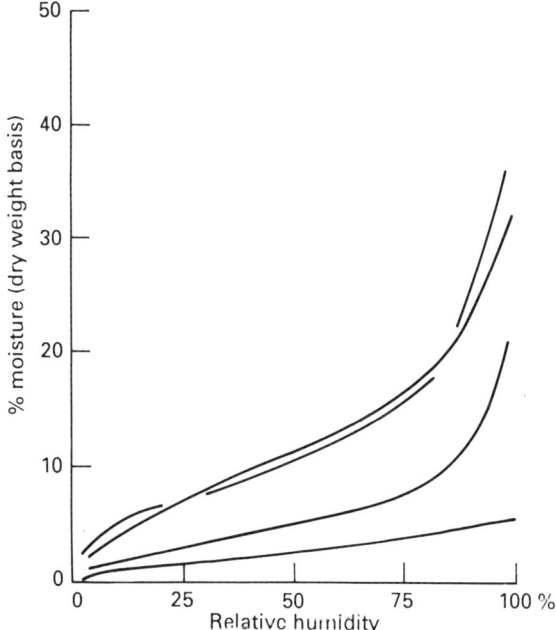

Figure 7.7 Sorption isotherm for textile fibres

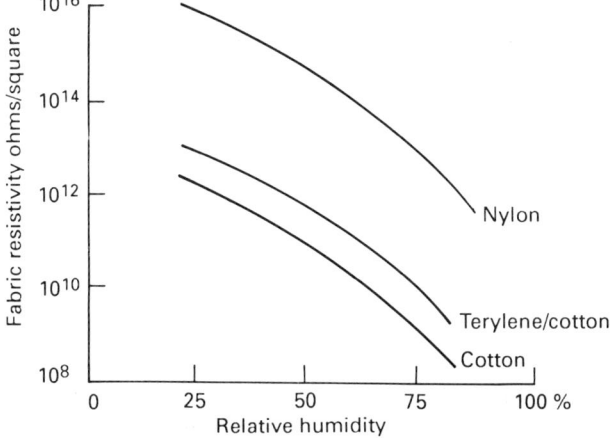

Figure 7.8 Fabric resistivity as a function of relative humidity

fibres. The electrical resistivities of different materials at different relative humidities are given in *Figure 7.8*. While non-absorbent fibres such as nylon have a high resistance, there is no simple relationship between actual water content and resistance for the different absorbent fibres.

7.3 Practical generation of charge

The most usual way of generating an electrostatic charge is to walk on a floor of high electrical resistance. The body voltage will rise quickly after each step at first and then after many paces reach an alternating high and lower fluctuation in time with the footsteps. An example taken from an electrostatic carpet test is given in *Figure 7.9*[12].

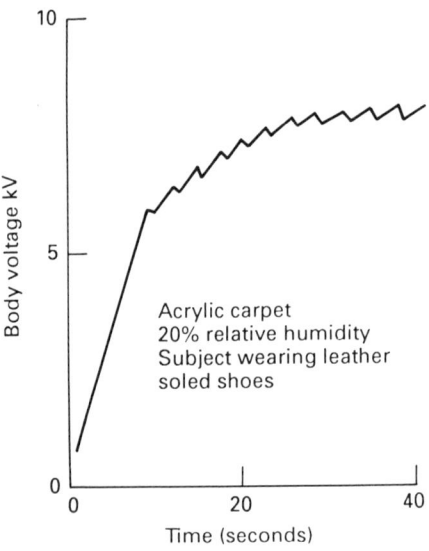

Acrylic carpet
20% relative humidity
Subject wearing leather
soled shoes

Figure 7.9 Illustrative build-up of body voltage when walking

The voltage generated is influenced by behavioural factors and physical factors. The behavioural factors are distance walked and height of step. The maximum voltage can take many steps to achieve and in small offices this is much less likely to happen than in the large open-plan spaces. Higher steps also result in higher voltages, and a 150-mm-high step can induce 50% greater body voltage than that generated by the same person only lifting his feet to a height of 75 mm[1].

The physical factors which affect the body voltage are the type of shoe material and the carpet characteristics. The most important carpet characteristic is its electrical resistance, which always varies with ambient relative humidity. The electrical resistance increases with decreasing relative humidity. All carpets have higher resistivity in dryer conditions but the absolute value of resistance is influenced by the weave, the thickness, the mixture of materials used in the carpet pile and backing, and the type of underlay. It is not possible to predict the electrostatic performance of a carpet and therefore there are a range of empirical walking tests designed to assess the body voltage which can be produced during a simulated walk. The magnitude and polarity of the voltage are also influenced by the different combinations of shoe sole materials and carpet fibre. This influence is shown in *Table 7.1*.

TABLE 7.1 Effect of shoe sole on static generation at 20% r.h. kV

Shoe sole	Carpet type				
	Wool	*Nylon*	*Acrylic*	*Polyester*	*Polypropylene*
Neolite	−17.0	−16.0	+3.0	+7.0	+8.0
Oak leather	−16.0	−15.0	+12.0	+8.0	+12.0
Chrome leather	−10.0	−13.0	+4.0	+9.0	+14.0

Chakravarti and Pontrelli (1976)[1].

The resistivity of the carpet can be measured in either the vertical or horizontal plane. The unit is the resistance in ohms of a square of the material. In practice there is a very good relationship between the vertical and horizontal resistivities, and the values are numerically similar[16]. The body voltage which is measured in an arbitrary walking test is approximately related to the carpet resistivity. This relationship tested for 27 different carpets is illustrated in *Figure 7.10*.

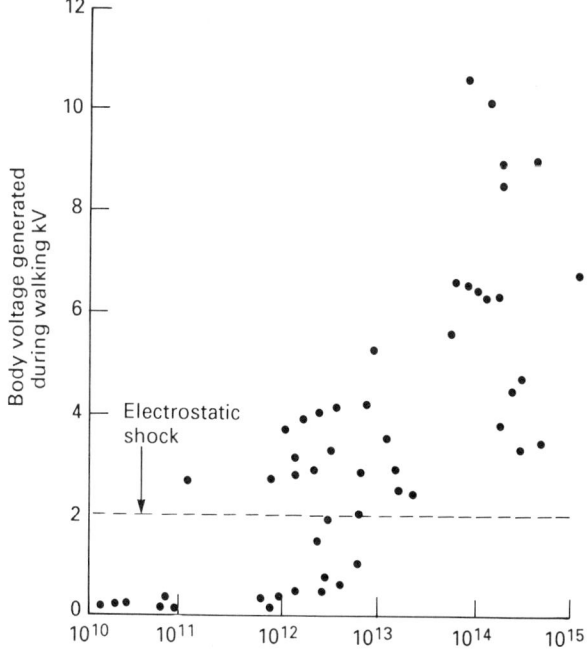

Figure 7.10 Relationship between maximum body voltage during walking and the vertical resistivity of carpets

The maximum voltage generated by walking on a carpet tends to be negligible above relative humidities of 60% for all carpets. The voltage rises with decreasing relative humidity, reaching a maximum around 10% r.h. Some carpets generated their maximum voltages at 20–25% r.h. and then diminished slightly at lower humidities. Both woollen and nylon carpets could generate between 13 and 17 kV body voltage in dry conditions. In extreme cases a nylon carpet can give rise to sufficient voltage to give a threshold electrostatic shock (2000 V) at a relative humidity of 50%. However, the majority of nylon carpets behave similarly to those made of wool. Acrylic carpets are satisfactory down to 45% r.h. in extreme cases and down to 40% for most.

In general there is only a small difference between carpets made of artificial fibres and those made of natural fibres. However, if electrostatic shocks are reported at relative humidities around 50%, then the carpet will probably be of nylon, although not a typical one. A lower limit of 40% r.h. will avoid electrostatic shocks when walking on most carpets. At 35% r.h. electrostatic shocks would be expected when walking on most carpets. The relationship between body voltage and relative humidity for carpets in general is given in *Figure 7.11*. The dotted line shows the extreme values while the solid line illustrates average conditions.

One exceptional case is that of underfloor heating. The carpet temperature is then raised above the room air temperature. Since the

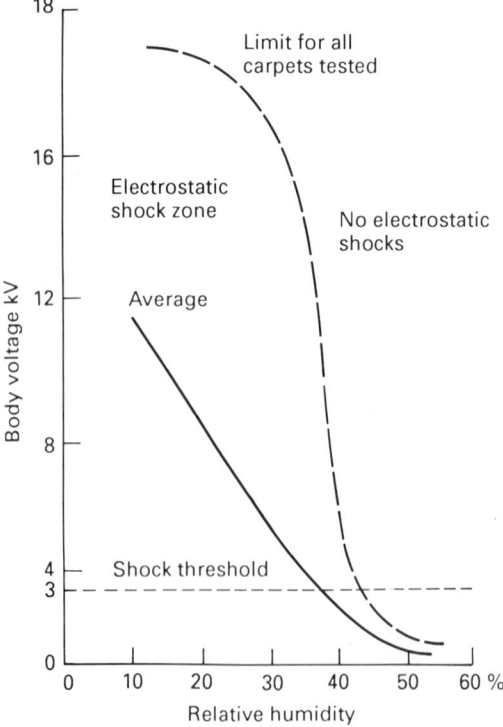

Figure 7.11 Relationship between maximum body voltage and relative humidity when walking on a carpet

moisture content of the air is constant, this higher temperature means that the relative humidity of the heated carpet is lower than that of the air. Since the electrical properties of the carpet are determined by the air in immediate contact with it, this means that these properties correspond with the lower relative humidity. This is illustrated in *Figure 7.12*. Point B is satisfactory for normal carpets, but to reach the same relative humidity C in the underfloor heated carpet at 25°C requires the relative humidity in the 20°C room to be 55%. Hence a typical elevation in carpet temperature of 5°C from 20°C to 25°C requires a relative humidity of 55% to equate to conditions at 40% r.h. and room air temperature.

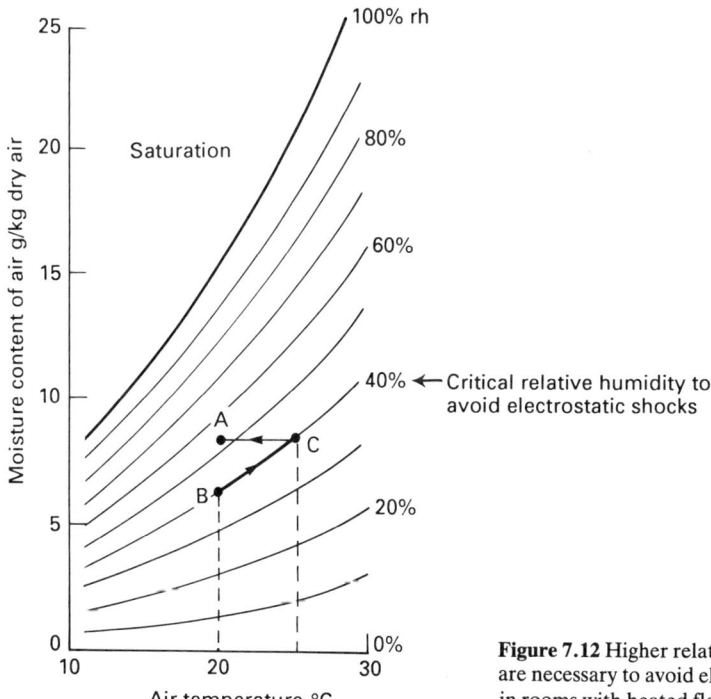

Figure 7.12 Higher relative humidities are necessary to avoid electrostatic shock in rooms with heated floors

Three other factors are used to modify the electrical properties of carpets. The first is addition of a hygroscopic material in the form of a shampoo. This humectant remained moist at lower relative humidities and delayed the onset of electrostatic shocks[17–19]. Unfortunately at high relative humidities the carpet could become damp and hence quickly soiled. More recent techniques permit the humectant to be absorbed within the fibre. This avoids the rapid soiling and slows down the removal of humectant during cleaning. In mild humidity conditions the addition of a particularly hygroscopic fibre has a similar effect. The addition of 5% of viscose fibre to a nylon carpet also significantly lowers the maximum voltage generated at a given relative humidity[20]. Hair underlay has a similar effect down to 35% r.h. but fails at lower values[21].

Other methods of improving the electrical conductivity of carpets which do not rely on moisture have also been tried with some success. Fine metallic wire, either copper or 8μm-diameter stainless wire, can be woven into the carpet yarn. Only 0.3% by weight is needed to make the carpet antistatic, although there is a tendency for this effect to decline with long use as the wires break and the electrical resistivity of the carpet rises[1,11]. Conducting nylon fibres woven into the carpet produce a similar effect. Such fibres, called epitropic, are bicomponent ones where the outer sheath of the fibre is enriched with sufficient carbon impregnation to enable it to become conducting. Only 0.1–0.2% of such fibres are needed in a carpet to reduce the body voltages to less than half of the voltages in an untreated carpet[3]. Conducting latex backing can also reduce the carpet resistance[1].

An electrostatic charge can also occur when walking on ordinary linoleum or tiled floor. The same principles apply, only instead of moisture influencing the electrical resistivity of the carpet, it affects the surface resistivity of the floor surface itself. Experiments on the maximum body voltage generated after one minute's walk show that voltages can be approximately 8 kV below 30% r.h., although at lower humidities the voltage declines. This relationship is illustrated in *Figure 7.13*[22].

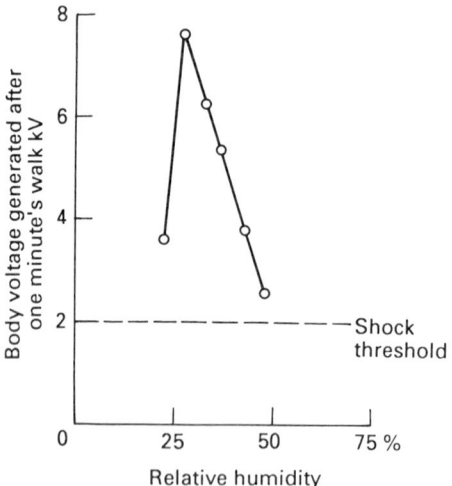

Figure 7.13 Body voltages generated after one minute's walk on linoleum

Any separation of dissimilar materials may result in charge generation. Such separation occurs when a person rises from a seat or when clothing is removed. In these cases the ambient relative humidity determines the leakage resistance and the nature and area of the two materials determine the magnitude of the effect. Experiments in hospitals on the body voltage generated when anaestheticians rise from their operating theatre stools showed that no charge occurred when the ambient relative humidity was above 45%. However, when the relative humidity fell below 36%, outer overalls of cotton and wool generated body voltages over 1000 V when the seat was upholstered in plastic. With a plain steel seat the voltage was 100 V.

Measurements of the body voltage created by rising out of a chair vary widely with the contact materials. Experiments on chairs covered in either Rexine (plasticized cellulose nitrate) or PVC (plasticized polyvinyl chloride) are shown in *Figure 7.14*. When the person wore a nylon boiler suit, the maximum voltage was over 3 kV and there was significant charge at the higher humidities when the two other materials only generated a very low potential[10].

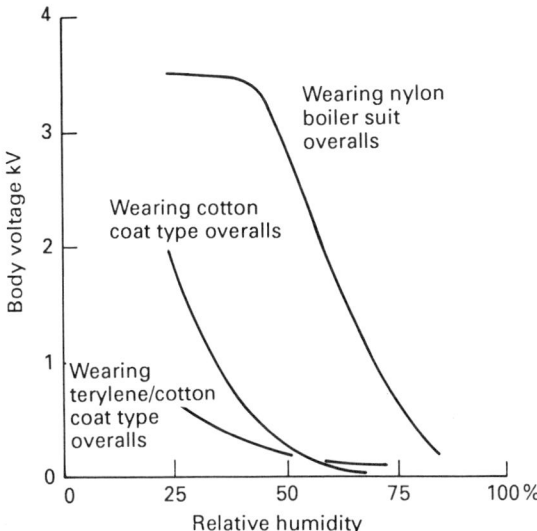

Figure 7.14 Body voltages when rising out of a chair

Voltages up to 5000 V were recorded on an operating table when the covering sheet was pulled off. Even at relative humidities of 53% voltages of 3500 V were recorded[10].

The highest charge potentials are associated with the rapid removal of outer clothing. Voltages over 50 kV were measured for materials made from cotton and for those of polyester/rayon mixtures. The highest voltages were recorded in the driest test conditions (~20% r.h.).

While cotton clothing can generate such voltages in very dry conditions, the maximum voltage falls rapidly with increasing relative humidity. The polyester/nylon mixture is still capable of generating over 20 kV even at 60% r.h. Clothing made with a proportion of electrically conducting fibres has a much lower voltage generated[23]. These results are illustrated in *Figure 7.15*.

Studies have shown that the electrostatic charge on arctic clothing assemblies increases with decreasing temperature. There is little hazard, providing the persons retain all their clothing. The hazard occurs when a person rapidly removes his outer jacket while working on a hazardous gas/air mixture[24, 25].

Electrostatic charges are generated by many other rolling or moving processes. The action of a vehicle tyre rolling on the road surface can

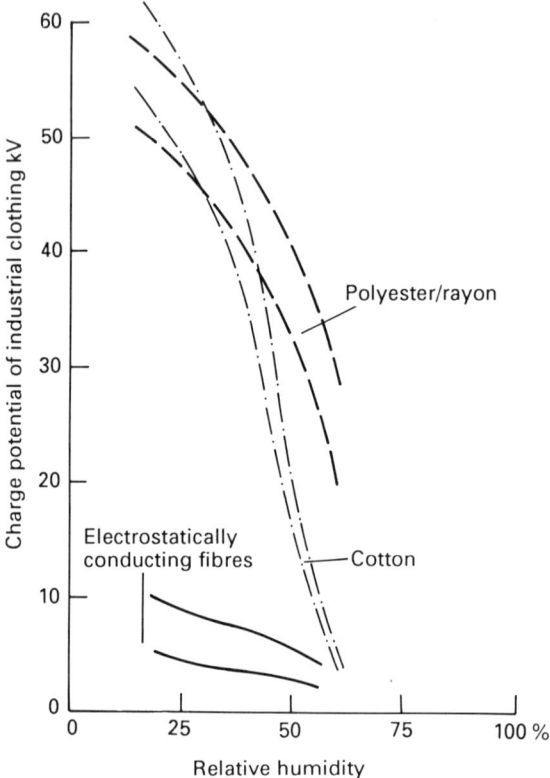

Figure 7.15 The charge potential when removing clothes

generate a charge potential of 100 kV on the vehicle. This not only creates an electrostatic shock hazard for passengers who touch the vehicle to enter it, but also can interfere with mobile communication equipment. The phenomenon can also generate ozone inside the tyre. This occurs if local roughnesses inside the tyre result in local voltage gradients exceeding 3000 V/mm. Ozone very rapidly attacks rubber and can create 'ozone puncture'.

Experimental relationships between the electrical resistance of the tyres and the vehicle speed on the voltage generated in the vehicle are illustrated in *Figure 7.16*[26].

Similar electrostatic problems can occur in the manufacturing industry where fibre is spun, printing processing handles paper or unrolls paper at speed, and in the film industry. Voltages of 10 kV are common and 40–60 kV are not unusual[23].

A special case of electrostatic charge generation is that associated with water sprays. Disintegration of high-speed water jets gives rise to a cloud of charged water droplets. The coarser drops all assume the same polarity while the finer droplets in the mist retain the opposite polarity. The larger drops precipitate more rapidly than the fine mist and this gravitational separation results in an electric field. This may lead to electrical discharges

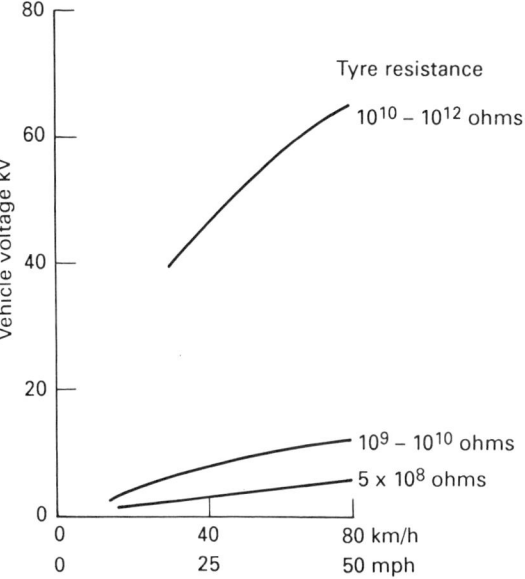

Figure 7.16 Vehicle voltage as a function of tyre resistance and vehicle speed

which could ignite any inflammable gases present[27]. Pneumatic pumping of fine powder in plastic pipes can also introduce electrostatic charging[28].

7.4 Personal sensitivity to electrostatic shock

The three factors which influence the magnitude of perceived shock when an electrostatically charged person is discharged to earth by a spark discharge are:

(1) The energy in the spark.
(2) The part and area of the body through which the discharge occurs.
(3) The personal sensitivity of the individual.

The electrical energy stored in a person is the product of his capacitance and voltage. The relationship is:

Energy stored E (joules) $= \frac{1}{2} \times$ Capacitance C (farads)
$\times$ [voltage V (volts)]^2

Illustrative values of capacitance are shown in *Figure 7.17*. The biggest variable is the size and thickness of the sole of the shoe. Large-area, very thin soles can have over 1000 pF capacitance. Sitting down can ensure a large capacitance. Typical values for a person would be a capacitance of 200 pF from hand to earth, and the energy stored at the conventional charge potential threshold value for shock at 2 kV is 0.4 mJ. The voltage has a more powerful influence than capacitance. The personal sensation, for different capacitances, is shown in *Figure 7.18*[28a].

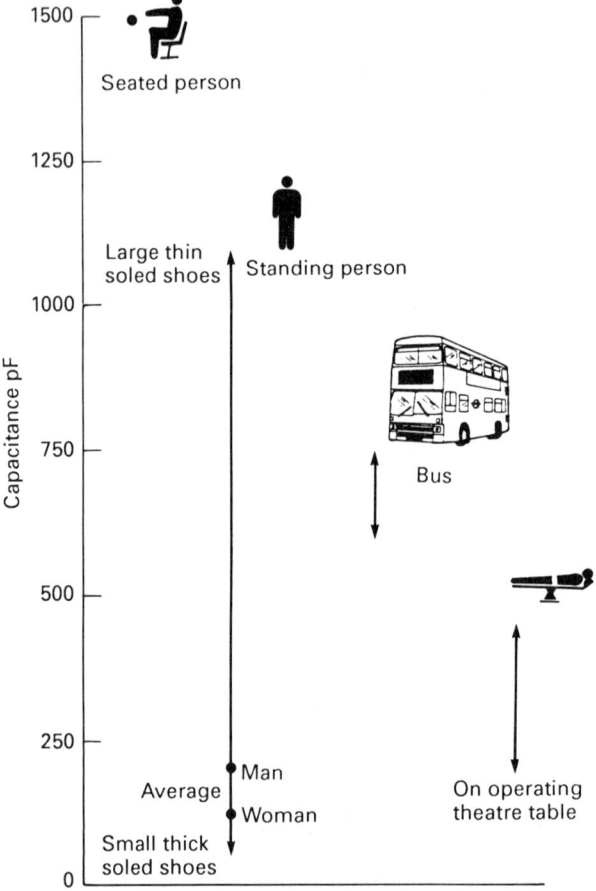

Figure 7.17 Illustrative range of capacitances

This charge is stored on the surface of the skin. The current flow is towards the point of discharge. The pain sensation is due to the local burning of skin at this point of discharge. If the current density can be reduced by drawing the discharge current over a wide skin area then the discharge will not be noticed. Discharges through the soles of shoes are not perceptible. Discharge through the tip of a metallic door key held firmly in the hand can be seen and heard but is unlikely to be felt.

Personal sensitivity to electrostatic shock is a relatively new concept. Early experimenters used themselves and their immediate colleagues, and their familiarity and scientific concentration made them relatively insensitive. The average population is more sensitive although there are wide differences between individuals. Assessments by large samples of up to 200 people show that there is close agreement between modern authors on the personal sensitivity to that voltage which they can detect when electrically discharged through the hand[1, 16, 29]. All people will detect a spark discharge through the hand when they are charged to a potential of

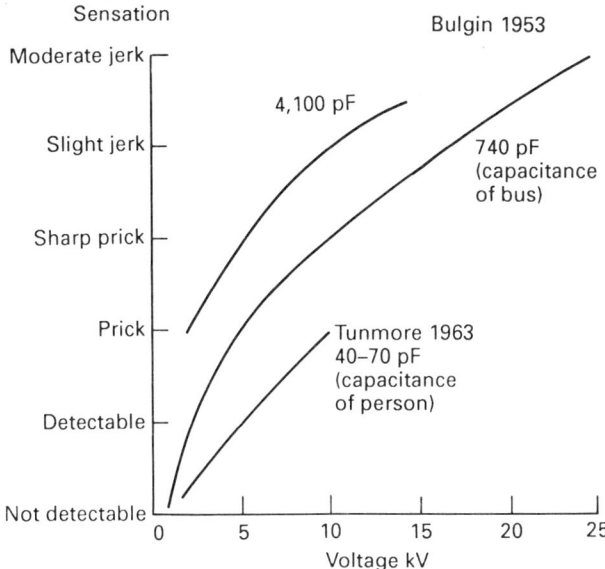

Figure 7.18 The influence of electrical capacitance upon sensation

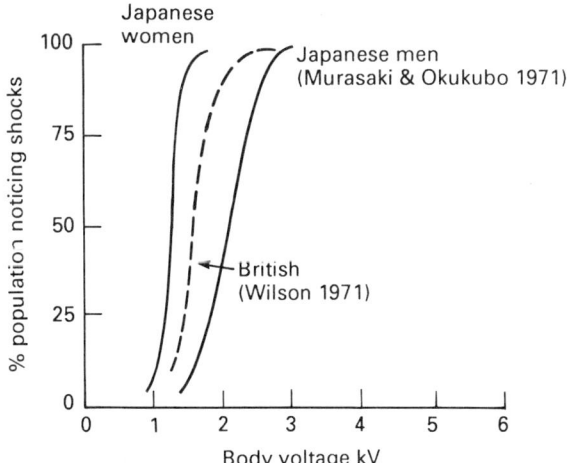

Figure 7.19 Population sensitivity to threshold detectability of shock

1–2½ kV. The cumulative distribution of population sensitivity is shown in *Figure 7.19*.

Objectionable shocks start at personal voltages of 25 kV, and at 7 kV most find the shocks unpleasant. Japanese women are the most sensitive group (*Figure 7.20*).

This recent and more refined analysis of personal sensitivity has led to a lowering of the acceptable body voltage from the 19-kV value used earlier[17, 26] to the present-day recommendations of 2.4 kV. This is the body voltage limit proposed for testing floor coverings for electrostatic shocks.

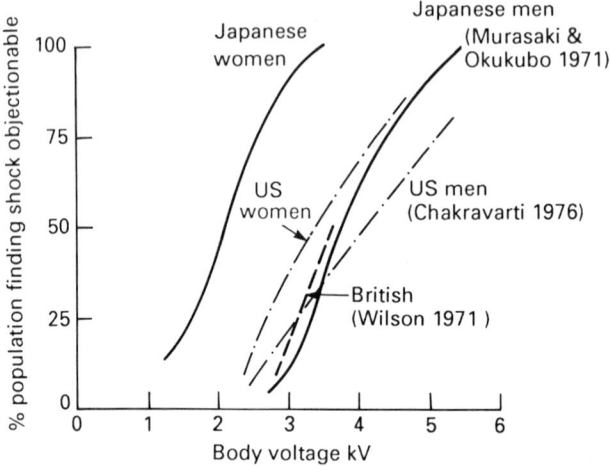

Figure 7.20 Population sensitivity to objectionable electrostatic shocks

In practice there is a large degree of randomness in the reported complaints of electrostatic shocks. It will depend upon the thickness and dampness of the shoe sole, its material, and the distance walked, in addition to the differences between people. One carefully recorded survey shows the kind of complaint record of electrostatic shocks in a large open-plan office (*Figure 7.21*)[30].

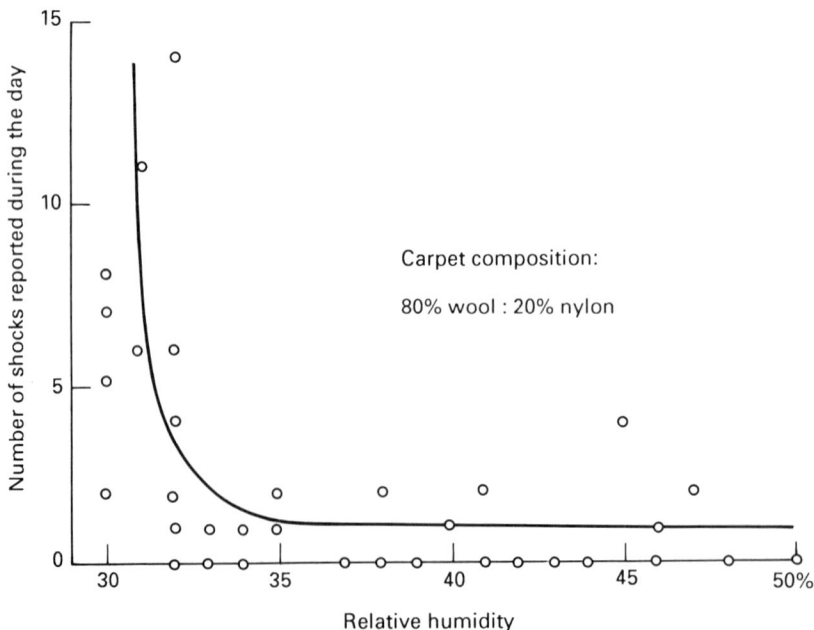

Figure 7.21 Office complaint study linking reports of electrostatic shock to the indoor r.h.

7.5 Safety criteria

Flammable gases, vapours and dusts can explode. The source of ignition could be an electrostatic discharge[9, 13, 23, 25, 26, 31–35]. In some circumstances the ignition energy required from the spark could be less than that required to create a detectable shock to the person. These circumstances are particularly dangerous because the person will not be aware of the potential danger.

Three conditions are necessary for a flammable gas to ignite[6]:

(1) The gas concentration is within certain limits of flammability. These limits depend upon the gas itself, the mixture and the physical conditions of pressure and temperature. Cyclopropane, a common hospital gas, will only ignite in air at concentrations between 2 and 10% by volume. These limits are considerably widened in an oxygen atmosphere to 2–63%.

(2) There is sufficient energy in the spark. This varies with the gas and is typically 0.03–0.3 mJ in air. The value can reduce to one-hundredth of this in an oxygen atmosphere. However, much more energy is needed as the gas/air mixture moves away from its ideal mixture which is stoichiometric.

(3) The spark length is sufficiently long or voltage sufficiently high. One long spark is much more effective than multiple small sparks discharging the same capacitor. Gas ignition by a multiple discharge to conducting fibres took 30 times more energy than the same discharge to a simple metallic sphere[6, 13, 14, 23].

Illustrative values of minimum ignition energy of gas mixtures are given in *Figure 7.22*. Not all the electrostatic energy stored on the person is available for the discharge. During discharge much of the stored energy is dissipated resistively in the skin and some heat is lost from the spark by electrode quenching. Experiments to measure the body energy needed in practice to ignite an optimum mixture of coal gas with air and natural gas with air showed that significantly more stored energy was needed for even an occasional detonation. For the coal gas mixture the person's stored energy had to be 100 times the minimum spark ignition energy of the gas mixture to achieve one explosion in 100 tests. The natural gas mixture required 60 times the minimum energy to achieve two explosions in 100 tests[6].

In normal atmospheres we can conclude that those circumstances where unpleasant electrostatic shocks occur are a serious potential danger to gas/air mixtures and precautions should be taken to eliminate the shocks. More recently electrostatic charge build-up has been demonstrated while high-pressure water jets are used to clean out large-volume oil vessels[27, 35]. In oxygen-enriched atmospheres of flammable gases, as in hospitals, imperceptible electrostatic discharges could cause explosions, and antistatic precautions are needed at all times.

The three conditions necessary for dust explosion are:

(1) The atmosphere must exceed a critical oxygen content. Most common flammable materials will explode in air. Organic dusts such as coal or rice will not explode below 16% oxygen content. Metallic dusts are more tolerant and can explode in oxygen concentrations down to 3%[36].

(2) The ignition energy must not be below a critical value. This varies widely from 10 mJ for phenolic resin dust to over 300 mJ for peanut dust.

(3) The dust concentration must exceed a minimum concentration. Coal dust must exceed a concentration of 35 g/m^3. Powdered metal dust is more sensitive and need only reach 20 g/m^3 for explosion. These are local concentrations around the ignition source. Dust explosions are unique in self-propagation. A small explosion can disturb a lot of dust into a cloud and create a much larger secondary explosion.

The ignition energies for dust explosions are summarized in *Figure 7.22* and in general are 100 times more than needed to initiate a gas explosion. There is little danger of dust explosions from personal sparks. The danger lies in charging up items of plant of large electrical capacitance when they create a powerful discharge.

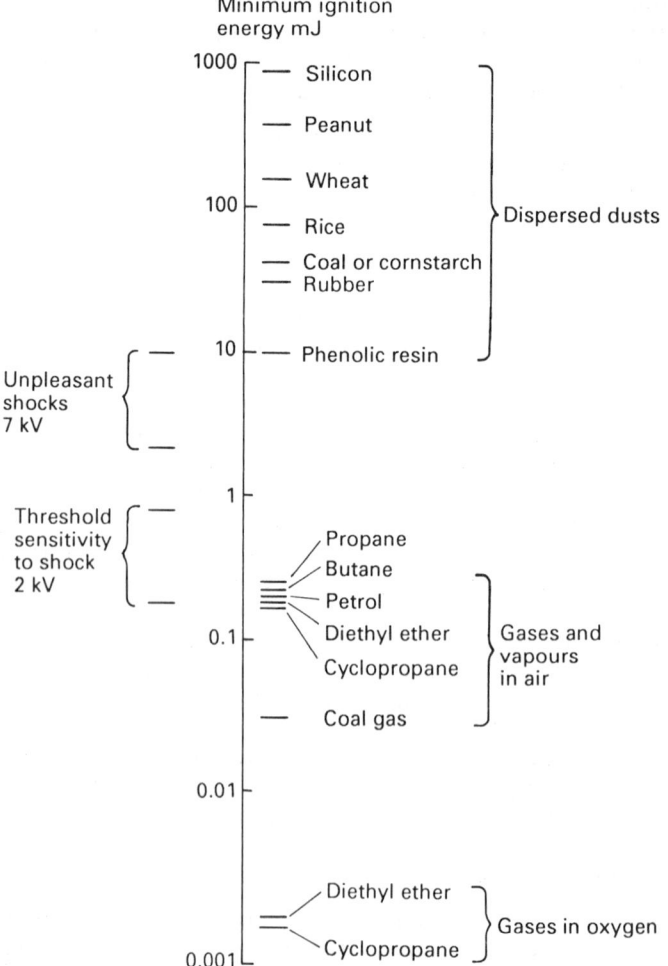

Figure 7.22 Minimum ignition energy for explosions

Some solid explosive materials are also easily ignited by sparks and very special precautions are needed at all times for these.

7.6 Guidelines

Moisture has a very large influence on the electrical resistance of fibres and surfaces. It can vary over a million times from dry to damp conditions. This makes it the most important single factor in determining the leakage resistance of an electrical charge and therefore on the ability of that charge to build up in potential and create a spark.

The energy stored through electrostatic charging is usually small (1 mJ) but at high voltage (~5 kV). The smallest practicable discharge is conventionally believed to be that across a gap of 0.01 mm which requires a voltage difference of 350 V[37]. Voltages up to 100 kV have been reported on machinery[26].

The guidelines are best examined in terms of decreasing hazard:

(1) *Safety*. There are certain areas where hazardous materials are handled and the smallest, even imperceptible, discharge could initiate an explosion. The activities include oxygen-enriched flammable gases in hospital operating suites. They also include the manufacture of explosive materials. In these areas the intention is to keep any voltage below 100 V and to humidify the atmosphere to 65% r.h. or more. There are other tasks which are potentially dangerous but not as sensitive as the first group. This would be dealing with refuelling operations, for example where there would be pockets of petrol/air mixtures which would be flammable or explosive. In these circumstances the ignition energy needed is normally above that which the people themselves would recognize as electrostatic. In these circumstances, whenever people report static problems then action has to be taken to eliminate the phenomenon.

(2) *Functional manufacturing problems*. While all production problems are unique to each factory, there is a consensus that electrostatic problems become troublesome when the relative humidity drops to 40% or below[23, 38]. Humidification is the simplest solution.

(3) *Personal electrostatic shocks*. This is most usually associated with walking across a floor and then touching a metallic contact such as a door handle or filing cabinet. It is rare to have this problem if the relative humidity remains above 40% unless underfloor heating is used. In this case the ambient relative humidity must be raised to 55% or above.

To summarize, for complete safety the relative humidity must be kept above 65%. Most processes and most people are not troubled by relative humidities down to 40%. When the relative humidity drops to 20% then most materials and processes display electrostatic problems.

Using British weather as an illustration, we can see how many days in the year the indoor relative humidity would be between these values. The data given in *Figure 7.23* assumes that outdoor air is taken and heated to 20°C.

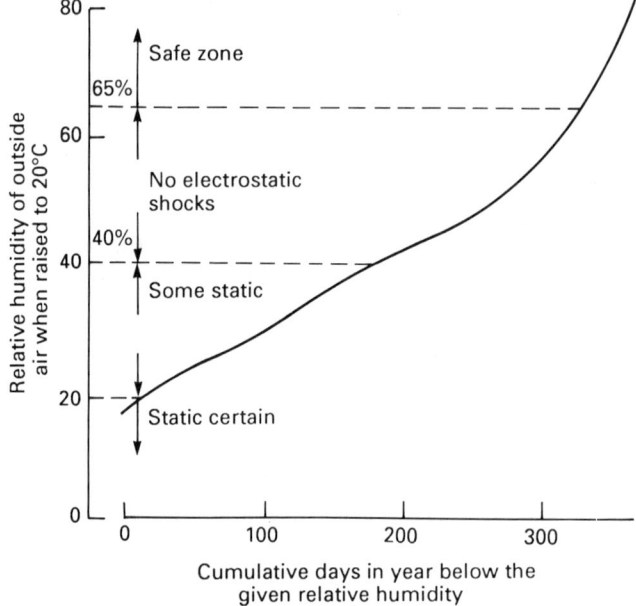

Figure 7.23 Cumulative days of the year below an arbitrary relative humidity

In practice there is usually some moisture added by the processes and the occupants of the buildings themselves. However, if electrostatic nuisances are recorded then the first step must be to record the actual relative humidity prevailing at the time.

References

1 Chakravarti, K. and Pontrelli, G.J. 'The measurement of carpet static'. *Text Res J,*, 129–134, February 1976
2 Corbett, R.P. and Hughes, J.F. 'Electrostatics – benefits and hazards'. *Engineering*, 1–8, July 1974
3 Ellis, V.S. 'Epitropics – third generation conductive fibres'. *Textile Manufacturer*, 19–23, July 1974
4 Eden H.F. 'Electrostatic nuisances and hazards', Chapter 18, pp. 424–440 in *Electrostatics and its Application* (ed. A.D. Moore). John Wiley, New York, 1973
5 Jowett, C.E. *Electrostatics in the Electronics Environment*. MacMillan, London, 1976
6 Wilson, N. 'The risk of fire or explosion due to static charges on textiles'. Shirley Institute Report for the Ministry of Defence, SCRDE 22/286/13/473: MOD Project No. A/78/CLO/47236/CB/CT/4B, 1973
7 Potter, A.E. and Baker, B.R. 'Static electricity in the Apollo spacecraft'. NASA Tech Note TND 5579, December 1969
8 Quinton, A. 'Safety procedures in operating theatres'. *Br J Phys*, Supplement No. 2, 92–94, 1953
9 Henry, P.S. 'Risk of ignition due to static on outer clothing'. *Inst Phys Static Elec Conf Series,* **11**, 212–221, 1971
10 Guest, P.G., Sikora, V.W. and Lewis, B. 'Static electricity in hospital operating suites'. US Bureau of Mines Report R14833, January 1952
11 Bajinski, G. and Lott, S.A. 'Electrostatic hazards from insulated operators'. Report S21: Australian Defence Scientific Service, Defence Standard Lab, Victoria, 1972

12 Cusack, J.A. 'Static control in carpets'. *Modern Textiles* Part I, **53**, 66–72, January 1972; Part 2, **53**, 25–29, February 1972
13 Wilson, N. 'The risk of fire or explosion due to static charges on textiles'. Shirley Institute Report 22/286/13: MOD Contract A/70/GEN/10004, May 1975
14 Wilson, N. 'The nature and incendivity of spark discharges from the body'. Shirley Institute Report 21/13/473: MOD Project A/78/CLO/47236/CB (CT) 4B, June 1976
15 Brundrett, G.W. 'A review of the factors influencing electrostatic shocks in offices'. *J Electrostatics*, **2**, 295–315, 1977
16 Wilson, N. 'The static behaviour of carpets'. *Text Inst Ind*, **10**, 235–239, 1972
17 Tunmore, B.G. 'Antistatic agents for use on carpets'. ERA Report 5017, September 1963
18 Hayek, M. 'Antistatic finishes for textiles'. *American Dyestuff Reporter*, **43**, 368–371, 1954
19 Valko, E.I., Resoro, G.C. and Ginilewicz, W. 'Elimination of static electricity from textiles by chemical finishing'. *Proc Am Assoc Text Chem Col*, **47**, 403–409, 1958
20 Anon. 'How KI developed a nylon staple for woven carpets'. *Textile Monthly*, 81–82, March 1972
21 Martin, D.H., Radford, R.D. and Lea, K.R. 'A shock free carpet system at 10% relative humidity'. *Modern Textiles*, 76–79, April 1971
22 Jonassen, N. *Statisk elektricitet og gulvbelaegninger*. Laboratet for Teknisk Fysik, Danmarks Tekniske Hojskole, 1978
23 Anon. *Prevention of electric charge on industrial clothing by means of conducting fibres*. Japanese Research Institute of Industrial Safety, 1971. Translated by ICI Fibres Division
24 Veghte and Millard. 'Accumulation of static electricity in arctic clothing'. US Air Force Report AAL-TDR-63-12, 1963
25 Quimby, W.S., Culbertson, T.L. and Mahley, H.S. 'Report of the task force on the potential hazards of synthetic clothing'. Texaco R&D Technical Report to the Static Electricity Committee of the American Petroleum Institute, 1968
26 Bulgin, D. 'Static electricity on rubber tyred vehicles'. *Br J Appl Phys*, Supplement 2, 583–587, 1953
27 Hughes, J.F. 'Electrostatic hazards in supertanker cleaning operations'. *Nature*, **235**(S338), 381–383, 1972
28 Hughes, J.F., Corbett, R.P., Bright, A.W. and Bailey, A.G. 'Explosion hazards and diagnostic techniques associated with powder handling in large silos'. *Inst Phys Conf Series* 27, Chapter 4, 264–275, 1975
28a Brundrett, G. W. 'A review of factors influencing electrostatic shocks in offices'. *J. of Electrostatics*, **2**, 295–315, 1977
29 Murasaki, N. and Okukubo, A. 'Some effects of static charges on the human body and possible remedies'. 3rd Shirley Institute International Seminar 'Textiles for comfort', Manchester, June 1971
30 Anon. 'Investigation of shocks by static electricity occurring at County Hall Island Block'. Greater London Council Bulletin No. 87, Item 6, pp. 6/1–6/2. July 1975
31 Fordham-Cooper, W. 'Practical estimation of electrostatic hazards'. *Br J Phys*, Supplement No. 2, 71–77, 1953
32 Thompson, R.E. 'Electrostatic safety in clothing'. *Fire Journal*, 15–16, November 1969
33 Crugnola, A.M. and Robinson, H.M. 'Measuring and predicting the generation of static electricity in military clothing'. US Army Textile Series Report No. 110, September 1959
34 Hughes, J.F., Bright, A.W., Makin, B. and Parker, I.F. 'A study of electrical discharges in a charged water aerosol'. *J Phys D: Appl Phys*, **6**, 966–976, 1973
35 Bright, A.W. and Hughes, J.F. 'Research on electrostatic hazards associated with tank washings in very large crude carriers'. *J Electrostatics*, **1**, 37–46, 1975
36 Anon. *Guide and Data 1985*. ASHRAE, Atlanta, 1985
37 Eichel, F.G. 'Electrostatics'. *Chem Eng*, **74**(6), 153–167, 1967
38 Ray, C.T. 'Influence of static electricity on paper printability'. Committee Assignment Report No. 62. Tech Assoc of Pulp and Paper Industry, Atlanta, 1975

Further reading

Beach, R. 'Industrial fires and explosions from electrostatic origin'. *Mech Eng*, **75**, 307–313, 1953
Bulgin, D. 'Factors in the design of an operating theatre free from electrostatic risks'. *Br J. Appl Phys*, Supplement No. 2, 87–91, 1953

Cohn, E.M. and Guest, P.G. 'Influence of humidity upon the resistivity of solid dielectrics and upon the dissipation of static electricity'. Bureau of Mines Information Circular 4492. June 1944

Hartgrove, E.H. 'Modifying static propensity'. *Modern Textiles*, SI (1), 41–44, 1970

Hearle, J.W.S. and Peters, R.H. *Moisture in Textiles*. The Textile Institute Manchester, Butterworths Scientific Publications, London, 1960

Hermack, F.L. 'Static electricity generated in fibrous materials'. NBS Report 4455, December 1955. US Department of Commerce

Hughes, J.F., Bright, A.W., Makin, B. and Parker, I.F. 'A study of electrical discharges in a charged water aerosol'. *J Phys D: Appl Phys,* **66**, 966–976, 1973

Jacobsen, M. 'Explosibility of agricultural dusts'. US Bureau of Mines Report S753. Boston 1957

Kisner, W.I. 'Causes and prevention of static markings on motion picture film'. *J Soc Motion Picture TV Eng,* **67**, 513–517, 1958

Klinkenburg, and van der Minne. *Electrostatics in the Petroleum Industry.* Elsevier, New York, 1958

Kunkel, W.B. 'The static electrification of dust particles on dispersion into a cloud'. *J Appl Phys,* **21**, 820–832, 1950

Long, J. 'Static can be eliminated when the relative humidities are right'. *Am Printer,* **124**(5), 31–35, 1947

McGuire, J.H. 'Fire danger from static electricity'. *Br Chem Eng,* 3(3), 136–140, March 1958

McLeod, T.S. and Johnson, G. 'Protection of data processing equipment against static electricity discharges'. *Electronics and Power,* 521–526, July 1978

Moore, A.D. (ed.). *Electrostatics and its Applications.* John Wiley & Sons, New York, 1973

Oxe, J. and Boschung, P. 'Prufung und Beurteilung des elektrostatischen Verhaltens textiles Bodenbelage im simultherten Begehtest., *Melliand Textilberichte,* **56**, 301–310, 1975. 'Testing and appraisal of the electrostatic behaviour of textile floor coverings using the simulated walking test'. Electricity Council Translation OA 1726, 1976

Ramer, E.M. and Richards, H.R. 'Correlation of the electrical resistivities of fabrics with their ability to develop and hold electrostatic charges'. *Text Res J,* 28–35, January 1968

Schiefer, H.F. and Hermach, F.L. 'Static electricity generated in fibrous materials'. NBS Report 4158. US Department of Commerce, June 1955

Sereda, P.J. and Feldman, R.F. 'Electrostatic charging of fabrics at various humidities'. *J Text Inst,* **55**, T288–298, 1964

Smith, J.C. 'Static electricity generated in fibrous materials'. NBS Report 4752. US Department of Commerce, June 1956

Thomson, R.E. 'Control of static electricity in textile materials'. *Safe Engineering,* 17–18, Oct/Nov 1969

Tobias, P.E. 'Moist air in sheet separation reduces static in offset press'. *The American Pressman,* 40–41, June 1958

Chapter 8

Aerosols, visibility and infections

8.1 Background

Sprayed water soon breaks up into droplets and when the droplets hit a solid surface they disintegrate into smaller droplets. Fine droplets ($10 \mu m$) stay suspended in air for a long time and are called an aerosol.

Sprays and aerosols have long been used in specialized drying techniques for the industrial, pharmaceutical and food industries. The techniques are now being applied to horticulture and farming to supply fertilizer and to distribute chemicals such as fungicides for plant protection. Sprays and aerosols are also becoming recognized as vectors for infection. In a mild form this could introduce humidifier fever from a dirty humidifier. A more serious example is legionnaires' disease, which is spread by the distribution and inhalation of a contaminated aerosol. Cloud physics are also an important area involving water aerosols and the ambient environment, which in turn influence the local climate.

Ambient relative humidity is the controlling factor which determines the physical changes which the droplet undergoes. It is also a major factor in determining the viability of most microorganisms in air and in dust.

Let us look at these two aspects, the physical and the microbiological.

8.2 Physical factors

8.2.1 Droplet size

When a body is dropped it accelerates rapidly at first under gravitational force. As the speed of fall increases so the aerodynamic drag forces increase and the acceleration slows. When the gravitational force equals the aerodynamic forces the body falls at a constant speed called the terminal velocity. The relationship between the size of a water droplet and its terminal velocity is illustrated in *Figure 8.1*[1]. Typical 'still air' conditions in an office are 0.1–0.2 m/s. Typical velocities in air conditioning ductwork vary between 2 and 10 m/s. Average outdoor velocities at 10 m height are 3–5 m/s with frequent peaks around 20 m/s[2].

Droplets exist over a very wide range of diameters. Haze droplets are

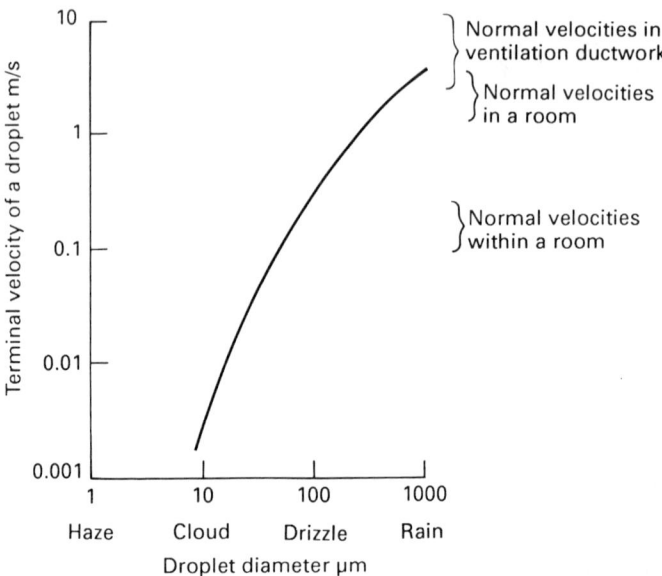

Figure 8.1 The terminal falling speed of a droplet as a function of diameter

the order of magnitude of 1 μm, fog or cloud droplets 10 μm, drizzle 100 μm and raindrops 1000 μm^3.

However, the water droplet tends to evaporate on exposure to air, at normal relative humidities. This has three effects:

(1) The droplet diameter decreases. This means that the terminal velocity reduces and the droplet remains suspended in the air for a longer period. The speed of reduction is illustrated in Figure 8.2[4] for a range of ambient relative humidities. This effect can also be observed in the plume length of the aerosol discharged into the atmosphere from an industrial cooling tower (*Figure 8.3*)[5].

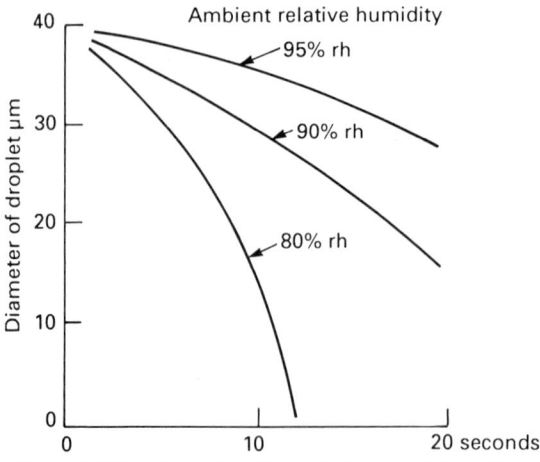

Figure 8.2 The effect of ambient relative humidity on the speed at which a droplet evaporates

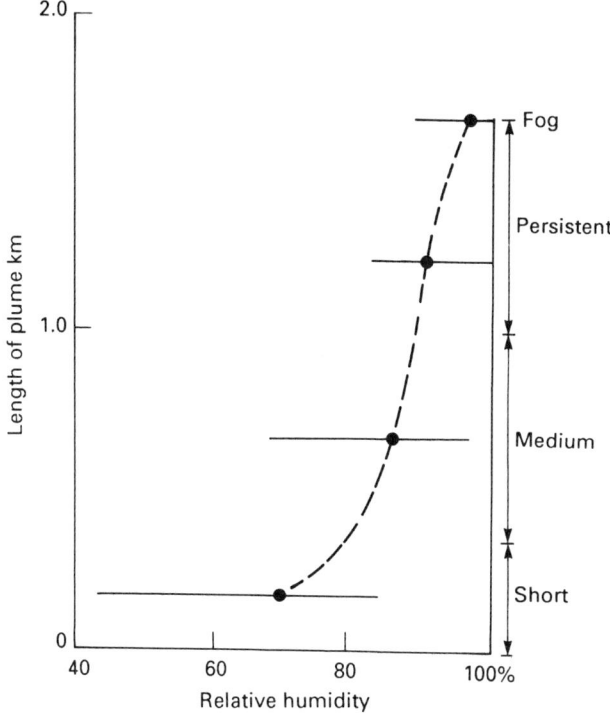

Figure 8.3 Plume length from industrial cooling towers varies with ambient relative humidity

(2) The concentration of chemicals increases. When the evaporation of the water occurs it leaves behind the dissolved chemical. These chemicals become progressively more concentrated as the evaporation proceeds, and if microorganisms are present may coat them with concentrated solute.

(3) The concentration of microorganisms within the droplet increases.

The evaporation process itself is a complex interaction of heat and mass transfer but a critical factor is the vapour pressure difference between water on the surface of the drop and that of the surrounding atmosphere. For a flat water surface the vapour pressure is a unique function of temperature. For a droplet this is complicated by three more factors:

(a) *The radius of the droplet*
Lord Kelvin demonstrated that the vapour pressure at the convex surface of a liquid would be higher than that of a plane surface. This effect is given by the following equation[3]:

$$\ln \frac{P_r}{p} = \frac{2o}{\rho R T r}$$

where P_r = equilibrium vapour pressure over a drop of radius r
p = equilibrium vapour pressure over a plane surface
o = surface tension of the liquid

R = Gas Constant
T = absolute temperature
ρ = density of the liquid

This increase in vapour pressure is small and for water droplets is only significant for droplets smaller than 10 μm. The magnitude of the increase is illustrated in *Figure 8.4*[6].

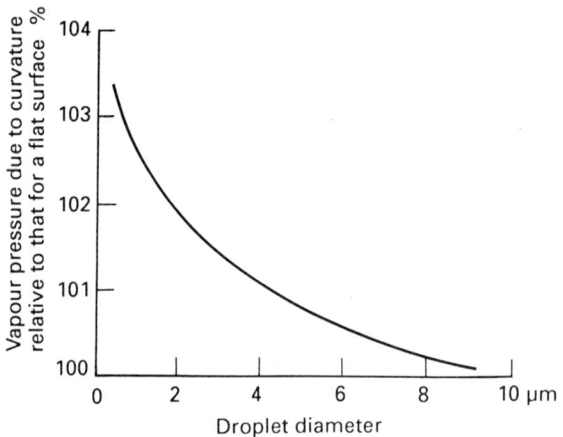

Figure 8.4 The effect of curvature on the partial vapour pressure of a water droplet

(b) *Impurities*
Any impurities, such as salts or acids, dissolved in the water decrease the saturation vapour pressure. The ideal relationship is expressed in Raoults Law[3] which is:

$$\frac{P_r}{p} = \frac{m}{M + m}$$

where m = number of moles of water
      M = number of moles of the solute (salt, acid, etc.)

In practice this relationship applies to ambient relative humidities approaching 100%. More generally, non-ideal properties must be determined and applied to Raoults Law. Changes in droplet size with changes in a relative humidity are strongly influenced by the deliquescent or hygroscopic nature of the chemical compound. Deliquescence is that property of compounds, primarily inorganic water-soluble salts such as sodium chloride, which causes them to form a solution when exposed to an atmosphere having a partial pressure greater than that of the saturated solution of their highest hydrates. Hygroscopic properties are attributed to a larger class of compounds which adsorb or absorb water vapour over a wide range of partial pressures by a variety of physical and/or chemical mechanisms[7].

If a beam of light is passed through an aerosol, some of the light will be scattered. The amount scattered increases as the aerosol droplet size grows. The scattering technique can be used to measure the effect of humidity on changing the droplet size in aerosols. Sea salt is a common coastal aerosol and the effect of relative humidity on light scattering is illustrated in *Figure 8.5*[7]. This demonstrates the characteristics of a deliquescent salt. The critical relative humidity at which a sodium chloride crystal of 0.04 μm diameter changes from a dry crystal to a solution droplet is 67%[8]. The experimental data are in good agreement with this.

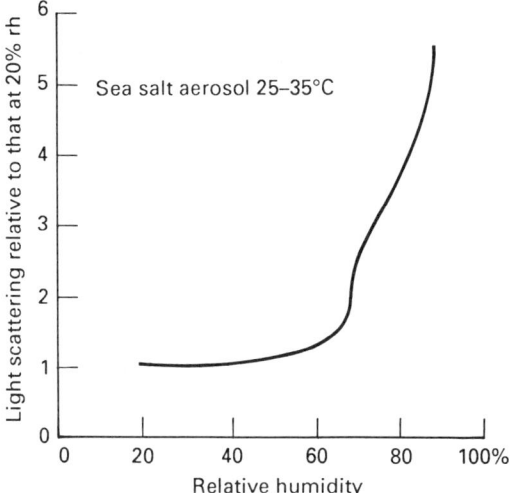

Figure 8.5 The degree of light scattering in an aerosol can be used to show the changes in droplet diameter which occur at different relative humidities

Hygroscopic compounds behave differently and do not exhibit a sharp transition step from solid to liquid. Experiments with a sulphuric acid aerosol show a smoother increase in droplet diameter with increasing relative humidity. This is illustrated in *Figure 8.6*[7].

Everyday atmospheric aerosols are a complex mixture of inactive particles, hygroscopic compounds and deliquescent nuclei. The overall

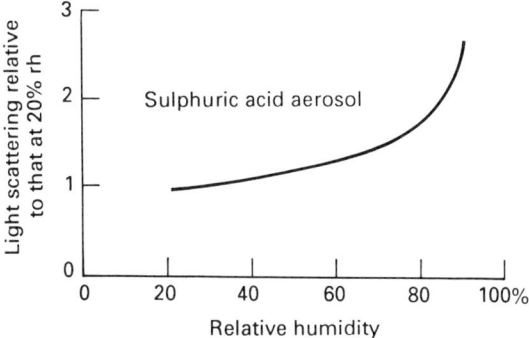

Figure 8.6 Characteristic changes in light scattering through a hygroscopic aerosol which reflect the growth of the droplet size

effect of relative humidity will depend upon the particular local composition. However, the influence is usually small for relative humidities below 70%.

(c) *Electrostatic charges*
Electrostatic charges on a droplet lower the partial vapour pressure. The relationship is expressed as[3]:

$$\ln \frac{p_r}{p} = - \frac{(nE)^2}{8\pi RT_r^4 \rho}$$

where n = number of unit electrostatic charges E

The influence of this factor is negligible except for very small droplets or droplets which are highly charged[3].

8.2.2 Visibility

The visibility through an atmosphere is determined by two kinds of particles. The first is the industrial haze associated with air pollution. Much of this is inactive and the obscuration it causes is principally the result of the physical mass of particles in the viewing direction[9]. The second component is the size, quantity and type of deliquescent or hygroscopic nuclei. These nuclei range in radius from 10^{-9} to 10^{-6} m, and are therefore comparable with or smaller than the wavelength of light. They tend to be invisible in dry conditions. Typical concentrations are $150 \times 10^9/m^3$ in cities and industrial areas, $50 \times 10^9/m^3$ in towns and $10 \times 10^9/m^3$ in the countryside[3]. The droplet size of such nuclei becomes important in terms of hindering visibility once the relative humidity exceeds 70%. While the overall effect of diminishing visibility with increasing relative humidity

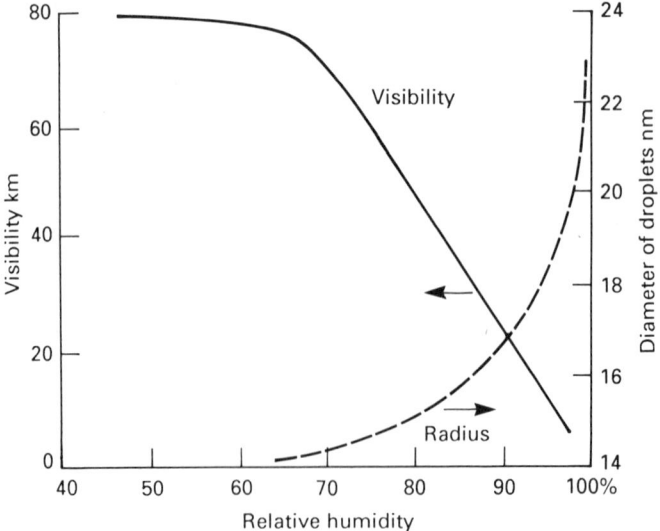

Figure 8.7 Average visual range and droplet diameter as a function of relative humidity

above 70% has been found by several investigators, the details differ from location to location according to the individual site conditions of quantity, size and composition of the condensation nuclei. One study of the relationship between visibility and relative humidity was based on an island in the North Sea. The measurements are presented in *Figure 8.7*[3], together with an estimate of the changing droplet size within such an aerosol. This result is compared with data from two other authors in *Figure 8.8*[10,11].

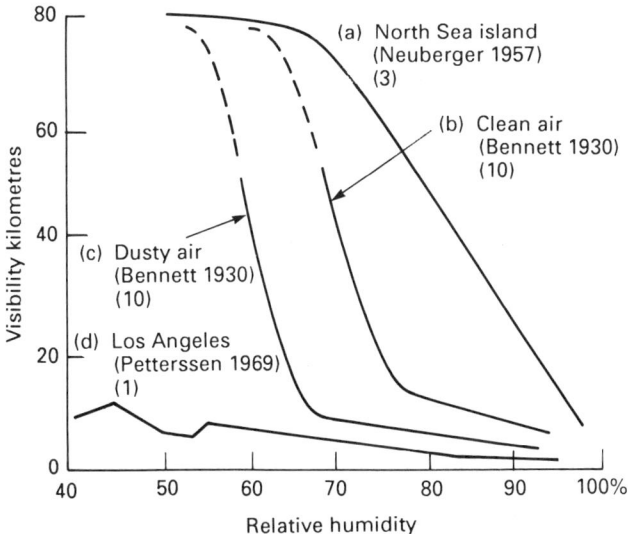

Figure 8.8 Visibility is influenced by the relative humidity

8.3 Microbiological factors

There are three stages in the life of a microorganism once it has been dispersed into the atmosphere through a waterborne aerosol:

(1) Surface drying of the water. This will occur with all aerosols when sprayed into a dry environment and will also result in the cells being coated with a concentrated solution of any solutes present in the water.

(2) Desiccation of the cell itself. The speed with which a cell loses its water varies widely from organism to organism, depending on the permeability of its cell wall.

(3) Survival in dust once the aerosol has eventually settled.

Most of the work in this field has been done on a small number of viruses and bacteria. While some general principles are emerging, the experimenters are agreed that most of the influencing factors, such as humidity, temperature, irradiance, the nature of the gaseous environment, the solutes and the cell history, all interact. There is also the belief that the laboratory experiments cannot properly reflect the complexity of the outdoor environment and that therefore an 'open-air' factor should be applied to laboratory predictions. This usually means that survival will be much less outdoors in daylight due to the trace contaminants in the atmosphere and the irradiance damage from daylight itself.

Infections are usually colonizations of the individual by either a virus or a bacterium. The initial dose required to induce infection has to be sufficient to overcome the natural body defence mechanism. The infective dose is therefore very different for different individuals, depending on their age, health and susceptibility, and on the novelty of the microorganism. Viruses and bacteria are the two most common infecting agents because they are the smallest and when suspended in an aerosol have the longest 'flying' time. Fungal moulds and yeasts are much larger in comparison and while both can cause infections and allergies, their growth characteristics are not included in this chapter.

Viruses are not strictly cells, because they have no metabolism of their own. They are very small, of the order of 10 nm, and can only be seen with an electron microscope. They are about one-hundredth of the size of a typical bacterium. They comprise a protein coat around a DNA strand of a few genes. They are parasites and kill host cells by attaching themselves to the healthy cell wall, introducing the DNA strand into the living cell. This takes over the cell's energy and materials to reproduce itself and destroy the cell. When the cell is destroyed the large number of viruses created repeat the attack on neighbouring cells. An illustration of the cycle is given in *Figure 8.9*[12].

The disease transmission route depends upon the particular virus. Some are transmitted through an aerosol. Colds and influenza are spread this way when an infected person sneezes. This is the route considered here, although direct waterborne transmission is possible as in polio, and insects can act as vectors as in mosquitoes transmitting yellow fever.

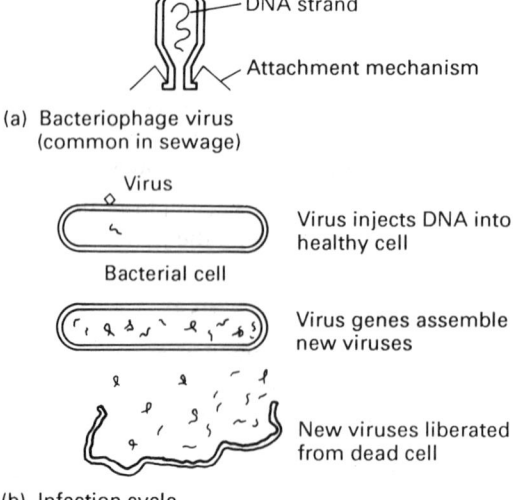

(a) Bacteriophage virus
    (common in sewage)

(b) Infection cycle

Figure 8.9 Multiplication of viruses

Experiments on viral infections have to be on living tissue, either in animals themselves, fertile eggs or special cell cultures. Early work on the influenza A virus showed a clear influence of ambient relative humidity. Mice were exposed for 15 minutes to an aerosol containing the virus. The droplet diameter was 1.5 µm based on mass modal diameter, that is 50% of the mass of the cloud was contained in droplets whose diameter was smaller. Both lung lesions and mortality were strongly influenced by the ambient relative humidity. Infectivity was highest at relative humidities below 40% and above 80% (*Figure 8.10*)[13]. This is consistent with the popular belief that respiratory problems become worse at low relative humidities, although planned field trials to resolve the question have found it difficult to provide satisfactory adequately matched controls[14–18].

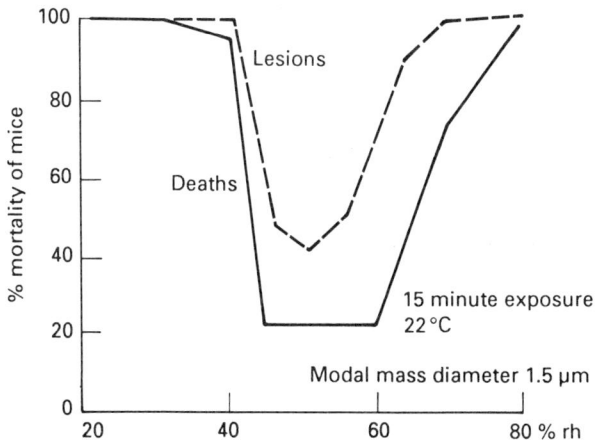

Figure 8.10 Mortality in mice exposed to an aerosol containing influenza virus in broth

Other experiments with influenza A virus agree with the enhanced survival at low relative humidities around 20–33% but show the progressive increase in cell death with increasing relative humidity up to 81% (*Figure 8.11*)[19]. *Vaccinia* virus died quickly at high relative humidities but survived best, at least for 15 hours, at 50% r.h. (*Figure 8.12*).

Experiments on the survival of the poliomyelitis virus in an aerosol showed the opposite effect of relative humidity to that of the influenza virus (*Figure 8.13*). The most lethal relative humidity to the virus was 50% and the viability of the cells was very low at 35% and 20% r.h. However, survival increased rapidly with relative humidities above 50%. There is no common pattern. Each virus appears to respond differently to humidity.

Bacteria are the smallest living cells, typicalling 1 µm in size. An illustration of a typical cell is given in *Figure 8.14*. The DNA strand, freely suspended in the jelly-like cytoplasm, gives instructions to the ribosomes to assemble protein chains from the nutrient surrounding them. The cell membrane is a very thin layer of protein and lipid and is freely permeable to water and gases. The cytoplasm itself is typically 80% water and all the metabolic activities occur in solution. Exposure to a drying atmosphere

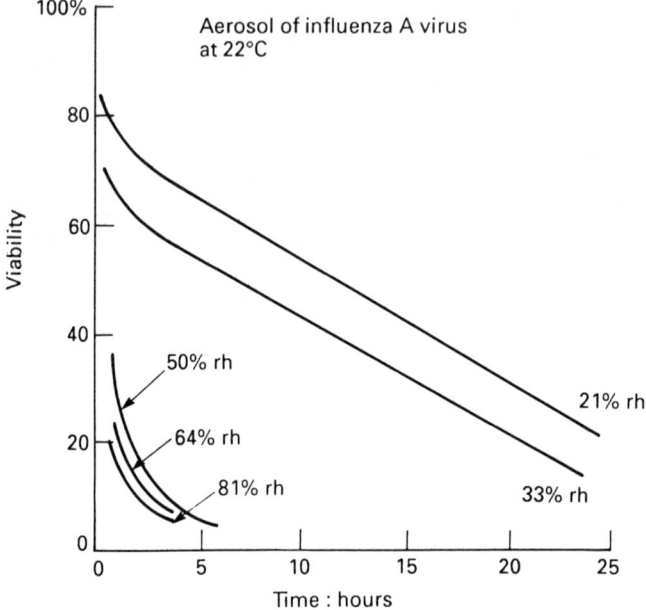

Figure 8.11 The change in viability of an aerosol of influenza A virus with time at a number of different relative humidities

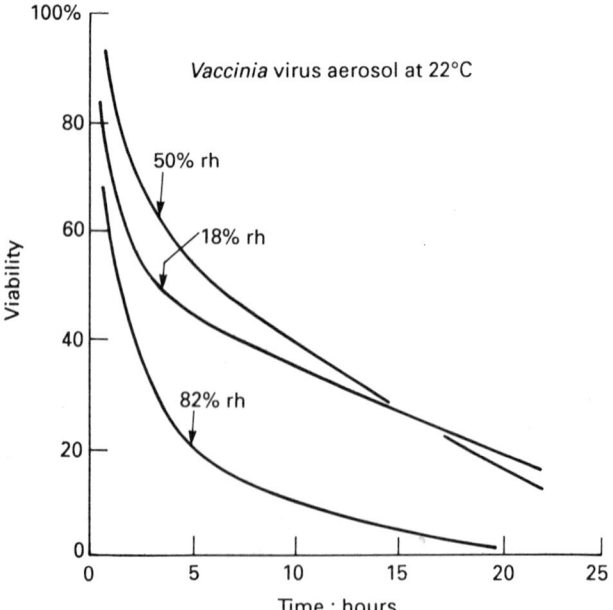

Figure 8.12 Viability of the *Vaccinia* virus at different ambient relative humidities

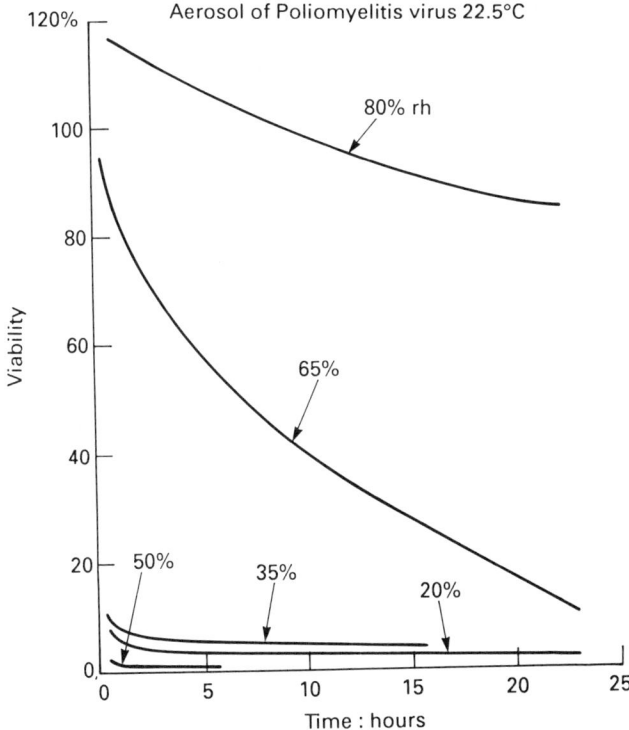

Figure 8.13 The change in viability of an aerosol of poliomyelitis virus with time at a number of different relative humidities

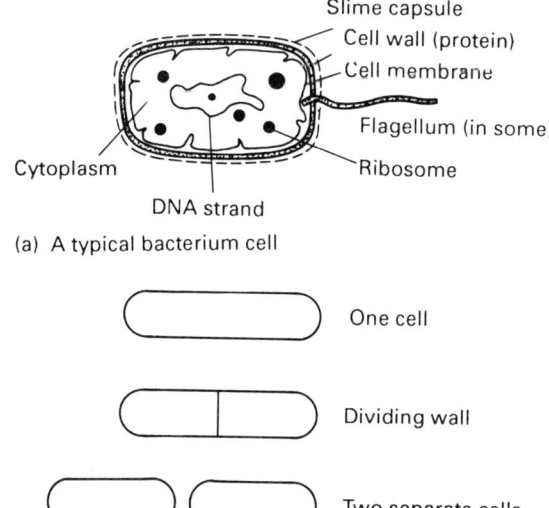

Slime capsule
Cell wall (protein)
Cell membrane
Flagellum (in some)
Cytoplasm
Ribosome
DNA strand

(a) A typical bacterium cell

One cell

Dividing wall

Two separate cells

(b) Binary multiplication of the bacterium cell

Figure 8.14 Multiplication of bacteria

would first remove any surface moisture, and then start to desiccate the cell itself. The speed of this desiccation would be related to both the cell and the atmosphere. The water within the cell would be bound with different bonding strengths, and release from the cell would also depend upon the permeability of the outer wall. The sensitivity of the cell to internal water content is also affected by the metabolic rate. In general, low metabolic rate cells survive better in aerosols. There is also the belief that there will be a critical cell moisture content at which the cell is particularly sensitive to damage from external pollutants.

Experiments with bacteria are more readily carried out than those with viruses. Culture techniques are used to sample an aerosol containing the bacteria at frequent intervals. This involves collecting a known volume of the aerosol at frequent intervals and culturing the microorganisms on a nutritional base medium. However, bacteria show the same idiosyncracies as the viruses. *Escherichia coli* in an aerosol survive well at 80% r.h. and die quickly at higher relative humidities. This is illustrated in *Figure 8.15*[20].

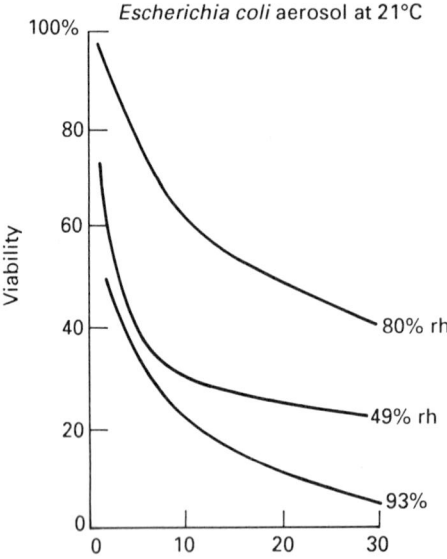

Figure 8.15 Viability of the bacteria *E. coli* at different relative humidities

An aerosol containing the bacteria *Legionella pneumophila*, the causative agent of legionnaires' disease, survives best in an atmosphere of 65% r.h. and dies quickly at 30% r.h. (*Figure 8.16*)[21].

Experiments with the bacterium pneumococcus aerosolized from a broth showed two distinct lethal processes[22]. The first was the most rapid and lasted for 5–20 minutes. The second death rate process proceeded more slowly and lasted for two hours or more. Both killing processes gave linear logarithmic survival plots. These are illustrated in *Figure 8.17*. The slope of these two lines is termed the logarithmic decay constant K. The short-term decay is K_1, the long-term decay K_2. When these survival experiments

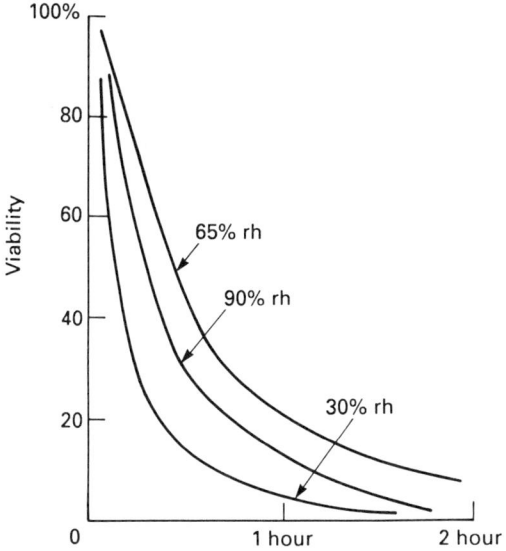

Figure 8.16 The viability of a *Legionella pneumophila* aerosol at different relative humidities

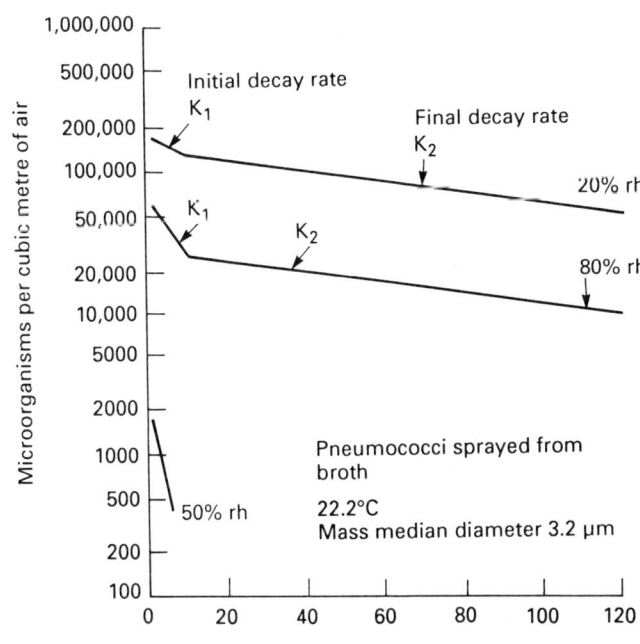

Time : minutes after introduction of microorganisms

Figure 8.17 Logarithmic plot of the survival of pneumococci sprayed from broth into atmospheres of various relative humidities

carried out over a wide range of relative humidities, it revealed a very narrow range of relative humidity in the vicinity of 50%, which is particularly lethal to this relatively delicate vegetative microorganism (*Figure 8.18*). Similar, albeit less dramatic, effects were observed in similar tests using *Staphylococcus albus* and *Streptococcus hemolyticus*.

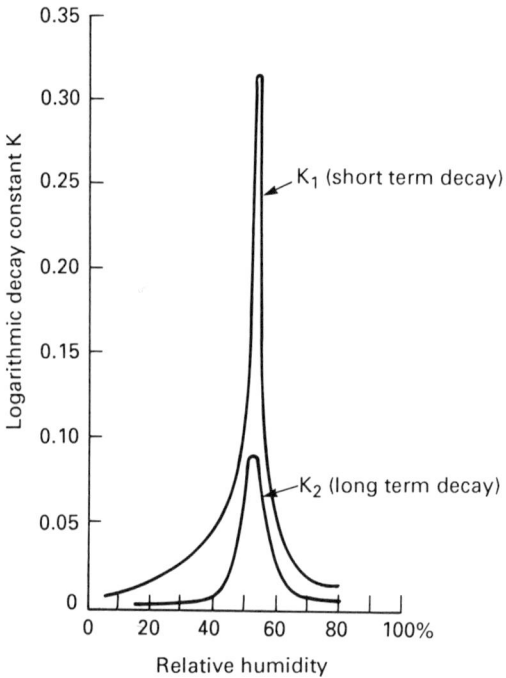

Figure 8.18 Slopes of the logarithmic survival curves for pneumococci sprayed from broth into atmospheres of various humidities

When these experiments were then repeated using an aerosol of the bacteria generated from distilled water rather than the broth, the sharp lethal effect of relative humidity at 50% disappeared. The pneumococcus was relatively insensitive to the ambient relative humidity except for a slight enhancement in survival for relative humidities below 20%. When the experiments were repeated using human saliva as the suspension medium, the survival curves returned to those of the broth, with a sharply defined lethal zone of relative humidity around 50%.

These experiments suggest that the concentrating effect of evaporation on solutes such as salt can lead to the cell being coated, interfering with its natural water balance. This then dehydrates the microorganism to the point where it becomes vulnerable. The lethal action of the salt probably involves denaturation of one or more essential enzyme systems.

These complicating factors, influenced by aerosol size, which determines the degree of chemical concentration, make it difficult to predict the influence of relative humidity precisely. The complication increases even further when irradiation is involved.

The bactericidal action of sunlight was recognized in 1877 and the lethal action of radiation on microorganisms is now well established. The relative effectiveness of wavelength in killing bacteria is illustrated in *Figure 8.19*[23]. In a general way this curve resembles the absorption curve for nucleic acid. This relationship is independent of the type of bacteria. It is most effective at 0.26 μm wavelength and practically zero beyond 0.32 μm.

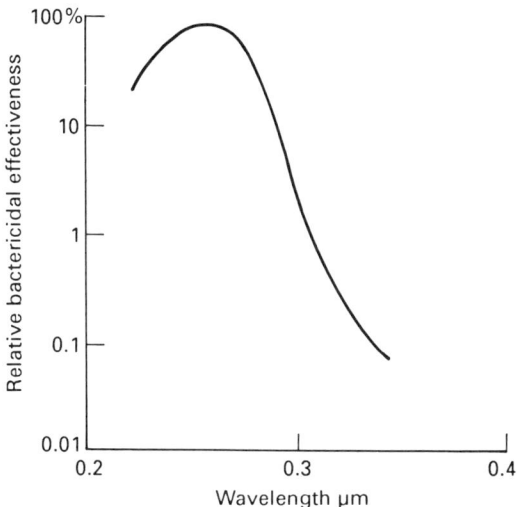

Figure 8.19 Bactericidal effectiveness varies with wavelength

The lethal effect of the radiation is a function of the energy received by the microorganisms[24]. Within wide limits the effects of weak radiation over a long time are similar to those of strong radiation over a short time. Time and irradiance are multiplied to give dose. Temperature appears to have little effect between 5 and 37°C.

Experiments designed to simulate sunlight explored the effect of irradiation and relative humidity on the death rate of the bacteria *Pasteurella tularensis* suspended both dry and in a wet aerosol. The sunlamps used a filter to eliminate wavelengths shorter than 0.305 μm[25].

The results for the wet disseminated aerosol showed a strong influence of both irradiance and relative humidity. The kill rate increased linearly with decreasing relative humidity and with increasing irradiance (*Figure 8.20*). Surprisingly, the experiments using dry dispersion of the bacteria showed a strong effect of irradiance but only a small effect of relative humidity. However, the dry particles consisted of an organic matrix comprising not only the bacterial cells but also milk solids and sugar.

There has been little research on the viability of bacteria in dust. The water adsorption isotherm for the bacteria *Serratia marcescens* is illustrated in *Figure 8.21*[26]. The cells will equilibrate with ambient relative humidity and take up or lose water. In dry conditions, below 20% r.h., the cells become desiccated and are relatively immune to damage. At very high humidities the cell is well endowed with water and prospers. At some

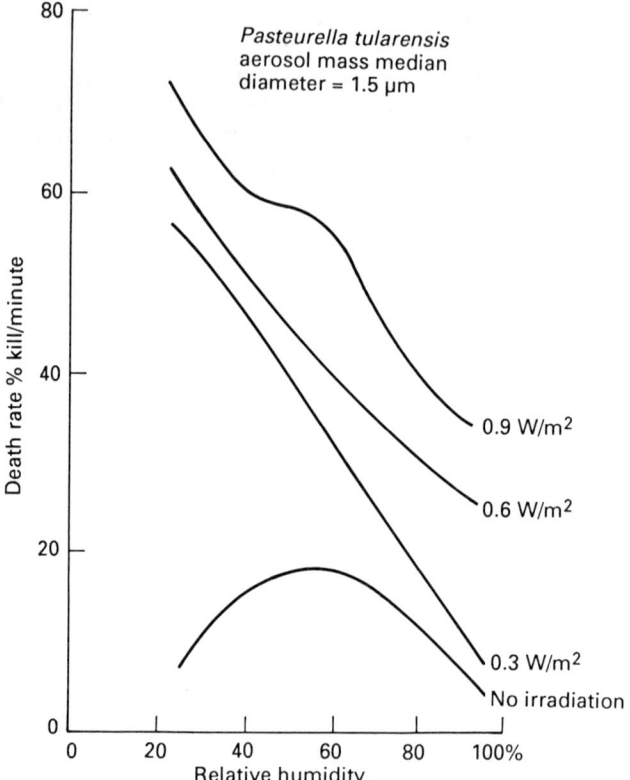

Figure 8.20 The influence of both radiation and relative humidity on killing the bacteria *Pasteurella tularensis*

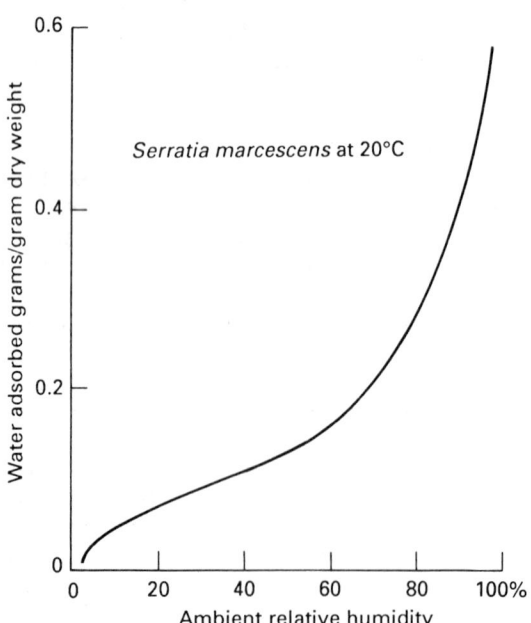

Figure 8.21 Water adsorption isotherm for the bacteria *Serratia marcescens*

in-between state there may be enough water present in the cell to permit some metabolic processes to continue but not all. The cell may be particularly susceptible to damage in this stressed state.

There is concern that the incidence of bacterial infections is increasing and that bacteria in dust may be a contributing factor. Outbreaks of infantile gastroenteritis have been reported from hospitals and traced to *Escherichia coli*. Some strains of the organism have been isolated from the air and dust samples from within hospitals. Nosocomial pseudomonal infections are becoming more commonly reported and *Pseudomonas aeruginosa* has been isolated from several hospital sites. Bedpans and urine bottles are often heavily contaminated. Cross-infections with *Salmonella* have also been reported[27].

Dust from hospitals which had been shown to contain streptococci was exposed to a range of controlled humidities for up to 35 days. The results showed a positive correlation between atmospheric relative humidity and the death rate of bacteria. This relationship is illustrated in *Figure 8.22*[28]. Survival was much higher for total bacteria at the lower humidities, and high relative humidities were particularly damaging to the *Staphylococcus aureus* and streptococcal bacteria.

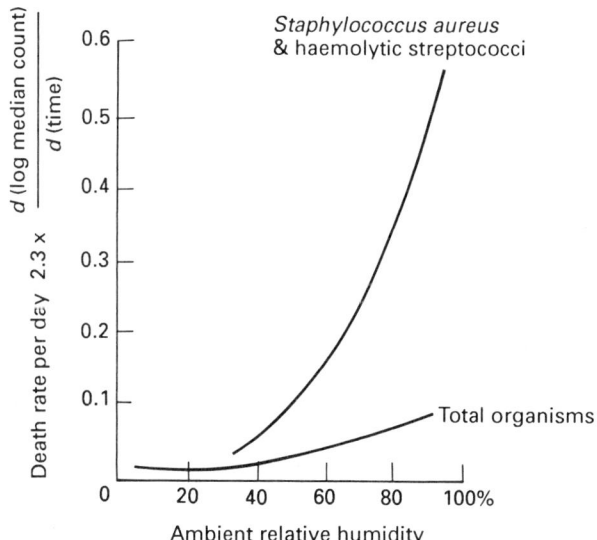

Figure 8.22 The death rate of bacteria in household dust

More recent experiments have explored the influence on survival of both relative humidity and the kind of surface on which the bacteria rest. Glass, metal and ceramic tile surfaces were used but there was very little difference between them. Relative humidity had a strong effect on both *Pseudomonas vulgaris* and *Proteus morgani*. Survival was very good at 11% r.h., 25°C, and viable organisms could be cultured after one week. The death rate increased rapidly with increasing relative humidity, and at 53% and 85% r.h. the surviving bacteria were less than one-thousandth of

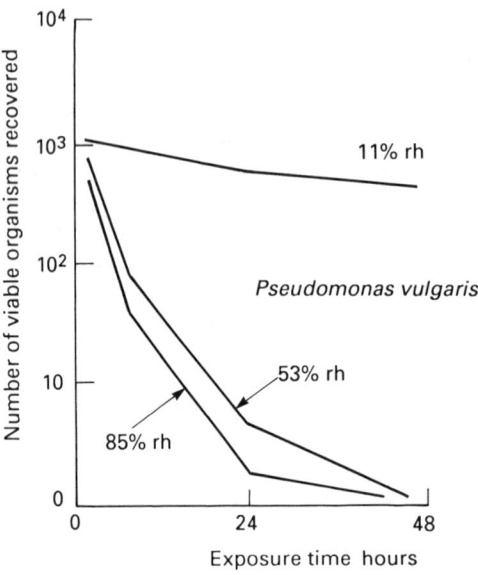

Figure 8.23 Survival rates for *Pseudomonas vulgaris* on metal surfaces at three different relative humidities at 25°C

the initial concentration after 48 hours. The results are illustrated in *Figures 8.23* and *8.24*[28].

Gases have long been used as disinfectants, and the early applications of burning sulphur to create sulphur dioxide and the use of chlorine to fumigate sickrooms are being replaced by chemicals less damaging to the room contents. Only in recent years has the importance of the ambient relative humidity become recognized. The effect is still a complex one because it interrelates with the state of hydration of the bacterial cells themselves, the presence of any protecting substances on the surface of the cell and the adsorption of moisture onto the solid surfaces on which the cells rest[29].

Illustrative properties of the most widely used disinfecting gases are listed in *Table 8.1*.

Ethylene oxide gas is the most widely used one, particularly for sterilizing plastic medical supplies. The gas is freely diffusible and penetrates many packaging materials such as paper, cellophane, cardboard and polyvinyl chloride plastic, although it is less readily diffusible through polythene. It can act as a moisture carrier through hydrophobic films which normally have very low moisture penetration. Moisture also allows the gas to pass through polar-type films which allow water to permeate readily but would normally impede the diffusion of the gas.

Under optimum laboratory conditions the maximum bactericidal effect is around 33%. However, the equally critical factor is the amount of moisture inside the cell. Since dehydrated cells are much more resistant in practice, the recommended operating conditions for disinfection are 30–50% r.h. and slightly higher if the contamination is heavy. The effectiveness is greater if the cells lie on moisture-absorbing surfaces.

TABLE 8.1 Properties of gaseous disinfectants

Gaseous disinfectant	Boiling point (°C)	Solubility in water	Sterilizing concentration (mg/litre)	Relative humidity requirements	Micro-biocidal activity	Application
Ethylene oxide	10	Complete	400–1000	Non-desiccated 30–50% Large loads 60%	Moderate	Sterilization of plastic
Propylene oxide	34	Good	800–2000	Non-desiccated 30–60%	Fair	Decontamination
Formaldehyde from formalin	90	Good	3–10	75%	Excellent	Surface sterilant for rooms
β-Propiolactone	162	Moderate	2–5	70%	Excellent	Surface sterilant for rooms
Methyl bromide	5	Slight	3500	30–50%	Poor	Decontamination

(Russell, 1976)[29].

Propylene oxide gas loses its effectiveness with increasing relative humidities, providing the bacterial cells are air-dried and not desiccated. In practice, environmental conditions of 30–50% r.h. are recommended. Higher relative humidities are recommended if the cells have been desiccated.

Formaldehyde gas is most effective at the higher relative humidities. The effectiveness is enhanced rapidly with increasing relative humidity up to 50% and thereafter at a lower rate. In practice an ambient of 75% r.h. is recommended.

The effectiveness of β-propiolactone increases with increasing relative humidity up to an optimum value of 75–85%. This is more directly related to the water content of the bacterial cell than to moisture in the air.

Sulphur dioxide is still used in commercial and industrial fumigation treatments, for example in greenhouses. Its effectiveness continues to advance with increasing ambient relative humidity. It is 20 times more effective at 96% r.h. than at 75% r.h. in killing the spores of *Botrytis cinerea*[30].

8.4 Conclusions

The physical influence of relative humidity on aerosols has its effect principally above 70%. At higher relative humidities the invisible condensation nuclei always present in the air act as centres on which moisture can condense or adsorb, and hence grow in size. Visibility is determined by the amount of light absorbed and scattered by droplets or particles present in the air. As the water droplets grow in size with increasing relative humidity, so the degree of light scattering increases and the visibility quickly reduces.

The persistence of vapour plumes is also strongly influenced by ambient relative humidity, particularly at the higher relative humidities. However, evaporation of water from a plume is also influenced by many other factors, such as wind turbulence, air and plume temperature, and solar irradiation. Plume lengths can, therefore, be quite short even in high ambient relative humidity.

Work on the influence of relative humidity on the infectivity of viruses and bacteria is both rare and valuable. Each microorganism appears to display its own characteristics. Relative humidity is an important factor for most microorganisms, although many other factors such as temperature are also important. The practical importance has only recently been recognized, for example, in the spread of legionnaires' disease when disseminated from a cooling tower. *Legionella* has an enhanced survival at high relative humidities, and hence is expected to have a much larger downwind infective risk zone at times of high ambient relative humidity.

In practice, survival of microorganisms is less outdoors compared with laboratory trials. This is attributed to an 'open-air' factor which is believed to be partly irradiation from sunshine and partly industrial contamination.

The survival of pathogens in dust has not yet received the attention it deserves but the data published to date show that viability is strongly influenced by the ambient relative humidity. Low relative humidities favoured the survival of those typical bacteria responsible for diseases caught in hospitals, namely *E. coli*, *Salmonella derby*, *Ps aeruginosa*, *Pr. vulgaris* and *Pr. morgani*. Humidification could benefit hospitals which operate in such dry conditions by reducing the incidence of nosocomial infections.

Relative humidity also has a critical role in determining the effectiveness of gaseous disinfectants. Its role differs for each of the gases and is modified by the task, the presence of surface contaminants on the cells, hydration of the cells, and the kind of surface on which the cells lie. Moisture can assist ethylene oxide gas to permeate polar-type films which would normally impede gaseous diffusion.

In general, ethylene oxide, propylene oxide and methyl bromide favour the lower ambient relative humidities for maximum effectiveness (30–50% r.h.) while formaldehyde and β-propiolactone are more effective above 70% r.h. Sulphur dioxide works best in an almost saturated atmosphere.

References

1 Petterssen, S. *Introduction to Meteorology*, 3rd edition. McGraw-Hill, New York, 1969
2 Lacy, R.E. *Climate and Building in Britain*. HMSO, London, 1977
3 Neuberger, A. *Introduction to Physical Meteorology*. College of Mineral Industries, Pennsylvania State University, 1957
4 Badenoch, Sir John. 'First report of the Committee of Inquiry into the outbreak of Legionnaires disease in Stafford in April 1985'. HMSO Cmnd 9772. June, 1986
5 Barber, F.R., Martin, A., Shepherd, J.G. and Spurr, G. 'The persistence of plumes from natural draught cooling towers'. *Atmospheric Environment,* **8**(4), 407–418, 1974

6 Day, J.A. and Sternes, G.L. *Climate and Weather.* Addison-Wesley Publishing Company, Massachusetts, USA, 1970
7 Covert, D.S., Charlson, R.J. and Ahlquist, N.C. 'A study of the relationship of chemical composition and humidity to light scattering by aerosols'. *J Appl Meteorol,* **11**, 968–976, 1972
8 Orr, G., Hurd, F.K. and Corbett, W.J. 'Aerosol size and relative humidity'. *J Colloid Sci,* **13**, 472–482, 1958
9 Charlson, R.J. 'Atmospheric visibility related to aerosol mass concentration'. *J Environ Sci Technol,* **3**, 913–917, 1969
10 Bennett, M.G. 'The physical conditions controlling visibility through the atmosphere'. *Q J R Meteorol Soc,* **56**(233), 1–30, 1930
11 Neiburger, M. and Wurtele, M.G. 'On the nature and size of particles in haze, fog and stratus of the Los Angeles region'. *Chem Rev,* **44**(2), 312–335, 1945
12 Ford-Robertson, J. *Revise Biology.* Charles Letts & Co. Ltd, London, 1981
13 Lester, W. 'The influence of relative humidity on the infectivity of airborne influenza A virus'. *J Exp Med,* **88**, 361–368, 1948
14 Lubart, J. 'The common cold and humidity imbalance'. *NY State J Med,* 816–819, March 1962
15 Sataloff, J. and Mendule, H. 'Humidity studies and respiratory infections in a public school'. *Clin Pediatr,* **2**(3), 119–121, March 1963
16 Sale, C.S. 'Reducing upper respiratory tract infections by allergy control and humidity control'. *Virginia Medical Monthly,* **96**, 368–371, 1969
17 Guberan, E., Dang, V.B. and Sweetnam, P.M. 'L'humidification de l'air des locaux previentelle les maladies respiratoires pendant l'hiver'. *Schweiz med Wschr,* **108**(22), 827–831, 1978
18 Green, G.H. 'The positive and negative effects of building humidification'. *ASHRAE,* **88**(I), 1049–1060, 1982
19 Harper, G.J. 'Airborne microorganisms – survival tests with four viruses'. *J Hyg (Cambridge),* **59**, 479–486, 1961
20 Anderson, J.D. 'Biochemical studies of lethal processes in aerosols of *E. coli*'. *J Gen Microbiol,* **45**, 303–313, 1966
21 Hambleton, P., Broster, M.G., Dennis, P.J., Henstridge, R., Fitzgeorge, R. and Conlan, J.W. 'Survival of virulent *Legionella pneumophila* in aerosols'. *J Hyg (Cambridge),* **90**, 451–460, 1983
22 Dunklin, E.W. and Puck, T.T. 'The lethal effect of relative humidity on airborne bacteria'. *J Exp Med,* **87**, 87–101, 1948
23 Koller, L.R. *Ultraviolet Radiation,* 2nd edition. John Wiley & Sons, New York, 1965
24 Wells, W.F. *Airborne Contagion and Air Hygiene. An Ecological Study of Droplet Infections.* Harvard University Press, Massachusetts, 1955
25 Beebe, J.M. 'Stability of disseminated aerosols of *Pasteurella tularensis* subjected to simulated solar radiations at various humidities'. *J Bacteriol,* **78**, 18–124, 1959
26 Bateman, J.B., Stevens, C.L., Mercer, W.B. and Carstensen, E.L. 'Relative humidity and the killing of bacteria: the variation of cellular water content with external relative humidity or osmolality'. *J Gen Microbiol,* **29**, 207–219, 1962
27 McDade, J.J., Hall, L.B. and Street, A.R. 'Survival of Gram-negative bacteria in the environment'. *Am J Hyg,* **80**, 192–204, 1964
28 Lidwell, O.M. and Lowbury, E.J. 'The survival of bacteria in dust: the effect of atmospheric humidity on the survival of bacteria in dust'. *J Hyg (Cambridge),* **48**, 21–17, 1950
29 Russell, A.D. 'Inactivation of non-sporing bacteria by gases', in *Inhibition and Inactivation of Vegetative Microbes* (eds F.A. Skinner and W.B. Hugo). Soc of Applied Bacteriology, Symp Series No. 5. Academic Press, London, 1976
30 Coney, H.M. and Uota, M. 'Effect of concentration, exposure time, temperature and relative humidity on the toxicity of sulphur dioxide to the spores of *Botrytis cinerea*'. *Phytopathology SI,* 818–819, 1961

Chapter 9

Horticulture

9.1 Background

Water is essential for all plants and 80% or more of the plant is water. This water serves several functions[1]:

(1) It provides the rigidity of the plant. Osmotic pressure is created within the plant cells by the diffusion of water into the roots. The pressurized cells form the turgid structure. Wilting occurs when there is insufficient water.

(2) It provides evaporative cooling of the leaves when they are tending to overheat.

(3) It provides the transport of mineral ions from the soil to the leaves where they are needed.

(4) It recirculates nutrients around the whole of the plant. This includes the sugars generated within the leaves by photosynthesis from atmospheric carbon dioxide and water. It also includes the circulation of the amino acids generated within the roots.

The overall movement of water is from the damp soil to the roots, up through the plant stem and finally to the leaves, where it is lost by evaporation. This process is termed transpiration. It is illustrated in *Figure 9.1*. Flows can be as high as 250 ml/hour for a tomato plant. The leaves are delicate membranes, usually only 10–15 cells thick, and are normally sponge-like in structure (*Figure 9.1*). The leaf itself is encased in a tough outer skin with the chlorophyll-rich cells just under the upper surface. Approximately 50% of the leaf volume is air space which is linked to the ambient air outside the leaf by tiny air valves called stomata. While the pore area of the stomata is only 0.5–2% of the leaf area, they are sufficiently abundant at 10–400/mm^2 that carbon dioxide diffusion and water vapour release are almost equivalent to free diffusion through the epidermis itself[2]. The stomata can open and close to regulate the gas and moisture exchanges. There are two banana-shaped guard cells at the stomata mouth. When these are turgid they arch into the surrounding subsidiary cells and leave a wide aperture. When relaxed and flaccid, they close. These guard cells normally open in daylight and close when dark.

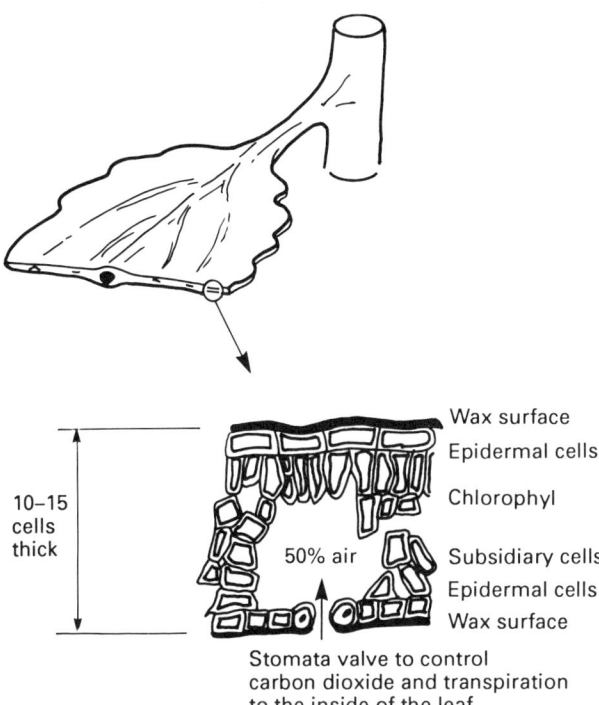

10–15
cells
thick

Wax surface
Epidermal cells

Chlorophyl

50% air

Subsidiary cells
Epidermal cells
Wax surface

Stomata valve to control
carbon dioxide and transpiration
to the inside of the leaf

Figure 9.1 Transpiration from a plant occurs through stomata, which are valves on the underside of the leaf

They also close in the light when the plant wilts and when there is a high concentration of carbon dioxide within the air space inside the leaf. They open when the leaf temperature rises so that additional moisture evaporation can provide some extra cooling.

It has long been known that the transpiration rate is influenced by atmospheric humidity. An illustrative example for a red kidney bean plant is given in *Figure 9.2*[3]. Low relative humidities result in high evaporative losses which result in an increase in the concentration of solute within the plant and hence an increase in water flow from the soil into the root system. However, there is little research on the topic because until recently the concept of treating humidity as an environmental variable was far too expensive. Two facts are changing this. The first is the development of crops in areas which have only recently been irrigated. The results are sometimes worse than expected because the ambient humidity is not the same as in the traditional growing areas. The second, and more relevant for greenhouse crops, is the introduction of low-cost efficient dehumidifiers. Early experiments have shown how they can protect crops from the potential disease problems associated with very high humidities. Recent years have, therefore, produced a whole new range of investigations into the influence of relative humidity on crop growth.

The overall conclusion is that, in general, higher relative humidities are better than lower ones, but there is a big variability between plants. There

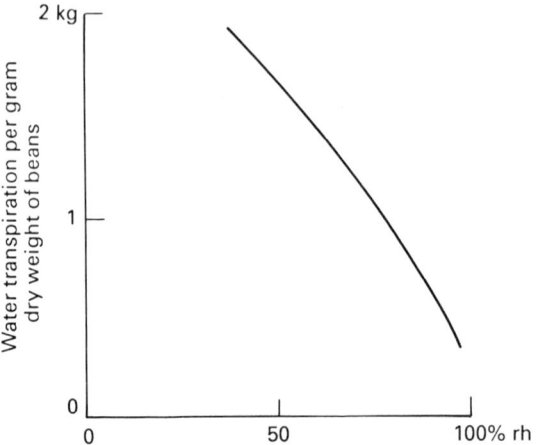

Figure 9.2 Changes in water transpiration with ambient relative humidity for a red kidney bean plant

are suggestions that flowers in particular respond best to those conditions which occur in their natural habitat[4]. Plants such as ferns enjoy wet conditions and welcome high relative humidities; cacti favour dry conditions. Suggestions for preferred plant relative humidities are listed in *Table 9.1*. In general, higher humidities are associated with bigger and thinner leaf growth but there is not a corresponding increase in photosynthesis overall. Very high humidities mean that there is little evaporation at the leaf and therefore little transport of mineral ions from soil to leaf[5–7]. This can lead to calcium deficiency and poor leaf growth which cannot be remedied by increasing the calcium ion concentration at the root. Very high humidities also increase the risk of disease.

TABLE 9.1 Preferred relative humidity for house plants (Condar, USA)

| | *Relative humidity* | |
30–45% r.h.	*45–60% r.h.*	*60% + r.h.*
Agapanthus	Achimenes	Azalea
Aspidistra	African violets	Camellia
Cacti	Amaryllis	Cyclamen
Century plant	Asparagus fern	Brake fern
Chenille plant	Begonia	
Chrysanthemums	Episcia	
Crocus	Fuschia	
Daffodils	Gardenia	
Hydrangea	Hyacinth	
Tulips	Spider plant	

9.2 Experimental evidence

One study of tomato plants examined how both the tomato and the leaf were damaged by disease in three types of environment[8]. The first type had no moisture control whatever, the second had a ventilation system which operated whenever the greenhouse humidity reached 90% r.h., and the third had a control set point of 75% r.h. *Botyritis* blemished the fruit and the effects of the different humidity control strategies are shown in *Figure 9.3*. The effect of high relative humidities was particularly pronounced in

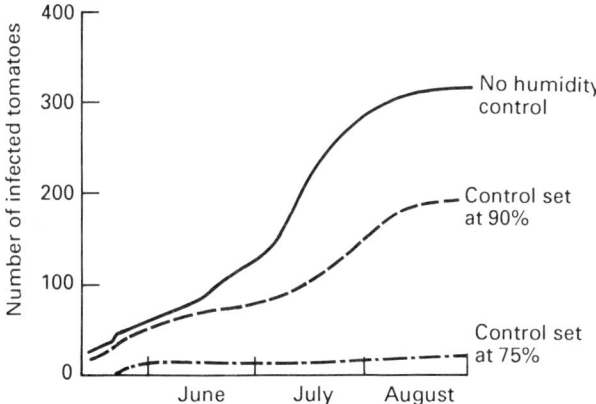

Figure 9.3 Effect of relative humidity control on the incidence of fruit blemishes caused by *Botyritis cinerea* on tomatoes

the month of July. Very few blemishes occurred when the control was set at 75% r.h. The mould *Cladosporium fulvium* did cause leaf lesions, and although this did not start until July it rapidly spread in the greenhouse without humidity control. One-quarter of the foliage was infected after six weeks. There were only a few lesions when the control was set at 90% r.h. and these only occurred at the end of the season. There were no lesions in the greenhouse with the moisture control set at 75% r.h. (*Figure 9.4*). Other researchers also studying tomato production found that those plants grown in a closed humid greenhouse lacked vigour in spring, were prone to high-temperature injury in summer and had poor fruit set in both seasons when compared with plants grown in a normally ventilated greenhouse[9].

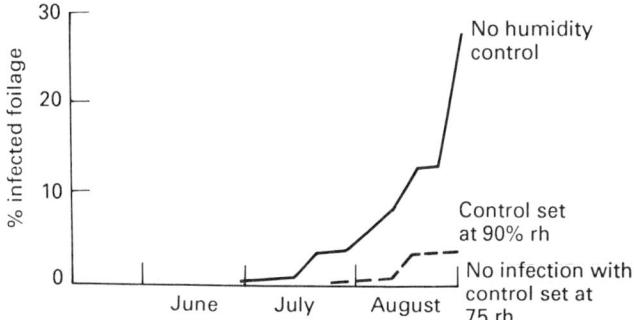

Figure 9.4 Effect of relative humidity control on the seasonal incidence of leaf lesions caused by *Cladosporium fulvium* on tomato plants

High humidity reduced the fruit quality in summer but not in spring. High humidity also induced uneven coloration, cracking and surface dullness. These defects reduced yields of marketable fruit by 56%. There were wide differences between the different plant varieties.

Experimental studies of the effect of different relative humidities on the seedling growth of potted plants showed wide differences between plants[10]. The three levels of relative humidity chosen were 40%, 65% and 90%. The three types of plant selected were representative of widely used bedding plants. They were all known to be responsive to experiments in controlled environments and they covered a range of photoperiodic responses. An ageratum was used because its photoperiodicity was independent of day length, a petunia which favours long days, and a marigold which is a short-day plant. For each test condition all the plants were in the same cabinet. Illustrations of the seedlings at the end of the 14-day experiment are given in *Figure 9.5*. The relationships between relative humidity and leaf area and fresh weight of the plant tops for both the ageratum and petunia are shown in *Figure 9.6*. Plant height and weight relationships with relative humidity for the marigold are given in *Figure 9.7*.

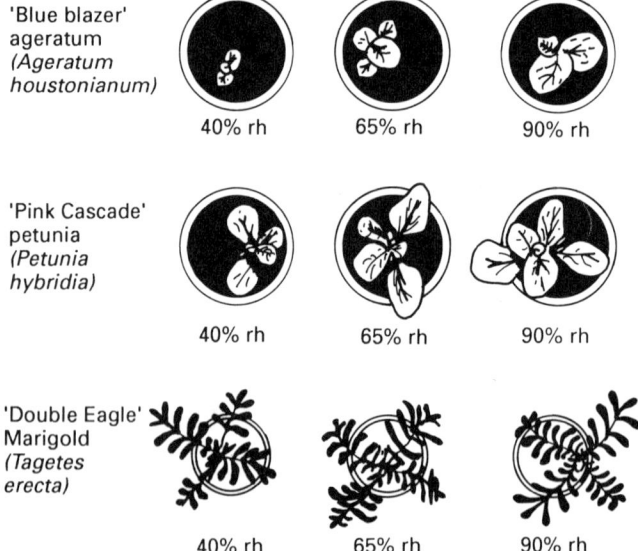

'Blue blazer' ageratum (*Ageratum houstonianum*)

40% rh   65% rh   90% rh

'Pink Cascade' petunia (*Petunia hybridia*)

40% rh   65% rh   90% rh

'Double Eagle' Marigold (*Tagetes erecta*)

40% rh   65% rh   90% rh

Figure 9.5 Illustrations of the effect of relative humidity on the growth of the seedlings of three annuals after 14 days

All plants benefited from increasing the relative humidity from 40% to 65%. Increasing the relative humidity further increased leaf area but not weight for the ageratum, but made little difference to the petunia and marigold. The experiments were carried out in two types of plant pot, clay and plastic, and there were unexplained differences which were attributed to the containers.

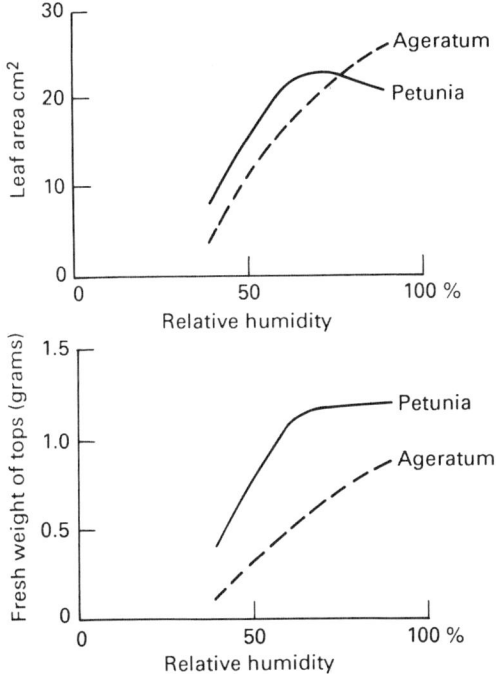

Figure 9.6 The effect of relative humidity on the growth of petunia and ageratum seedlings after 14 days in clay pots

Experiments on cotton plants show the complexity of the growth process[11]. The cotton plants grew taller with increasing relative humidity (*Figure 9.8*). The ratio of shoots on the plant to the amount of roots developed showed a large increase with increasing relative humidity, reaching a maximum value around 70% r.h. The cotton yield was very sensitive to humidity and reached an optimum at just over 50% r.h.

9.3 Physical principle

While the physical principles of evaporation have been known for almost two hundred years, it is only relatively recently that they have been applied to plants[12–17]. The important physical driving force for evaporation is the water vapour pressure difference between the water within the leaf and the water vapour pressure in the ambient air around the leaf. The actual rate of moisture flow will depend upon the opened pore area of the stomata on the underside of the leaf.

The relation may be expressed:

Transpiration rate m (g/m^2 per s) $= k_m (P_s - P_a)$

where $k_m =$ transpiration coefficient (g/m^2 per s per kPa)

(This transpiration coefficient will change as the stomata open and close and is increased by air flow around the plant)

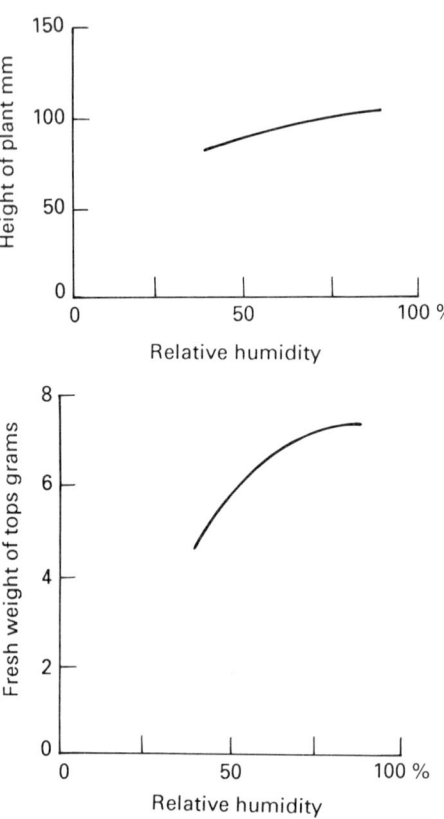

Figure 9.7 Effect of relative humidity on the growth of marigold seedlings grown in clay pots for 14 days

P_s = water vapour pressure at the evaporating surface (kPa)
P_a = water vapour pressure of the surrounding air (kPa)

$(P_s - P_a)$ is often termed the vapour pressure deficit.

P_s is usually calculated as:

$$\text{VPL} \times P_{sat}(T_s)$$

where VPL = vapour pressure lowering effect due to the presence of solutes in the leaf moisture; typical values are 0.98 or 0.99

$P_{sat}(T_s)$ = saturation vapour pressure for pure water at temperature T_s

When transpiration rates are expressed in terms of vapour pressure deficit for the leaves of the plant, a clear picture emerges[18]. The relationship is essentially linear but there are wide differences between plants. Illustrative results are given in *Figure 9.9* for those plants more associated with arid conditions.

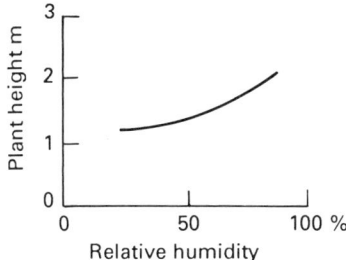

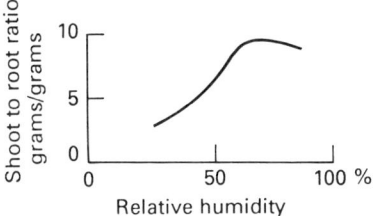

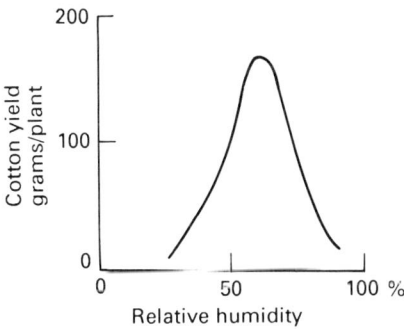

Figure 9.8 The effect of relative humidity on the growth of the cotton plant

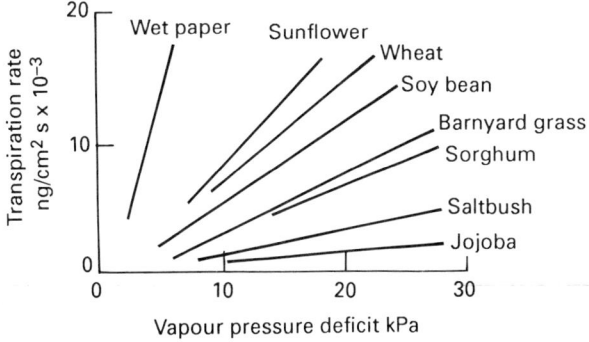

Figure 9.9 Transpiration rate for a range of plants at 26°C

9.4 Susceptibility of plants to pollution and change

The susceptibility of plants to air pollution is influenced by the environmental factors of temperature, air speed, light intensity, soil fertility, and soil and atmospheric moisture[19]. The supply of moisture both before and during exposure to pollutants is one of the most important factors. In general, conditions which minimize plant moisture stress increase plant susceptibility to leaf damage. High soil moisture content and high ambient relative humidities minimize plant stress.

Planned laboratory experiments have shown one of the complexities of moisture and pollution take-up. Red kidney bean plants were subjected to a range of concentrations of sulphur dioxide at both high and low relative humidities. The absorption of sulphur dioxide into the leaves was two or three times higher in the high ambient relative humidity ($\sim$77%) compared with the low relative humidity condition ($\sim$35%). There were suggestions that increasing pollution concentrations did not increase leaf absorption. Peak absorption occurred around 1 mg/m^3 of pollutant sulphur dioxide.

Similar experiments in an ozone-enriched atmosphere showed an even greater leaf absorption at high relative humidities. Some three to four times more pollutant was taken up at 70% compared with at 35% r.h. At high humidities there was a tendency for absorption to increase with pollutant concentration. At low relative humidities the higher pollutant concentrations were less absorbed.

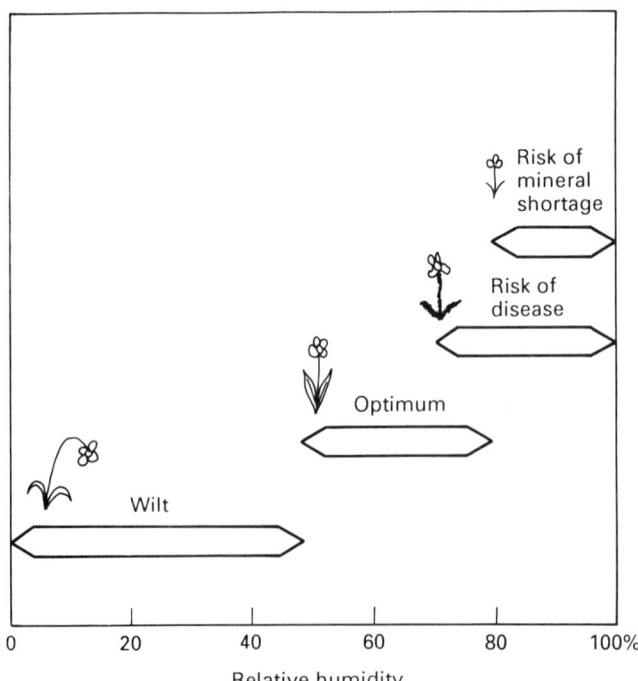

Figure 9.10 Humidity and horticulture

Low relative humidities are recommended when spraying fungicide on to chrysanthemums to combat petal blight[20]. However, sulphur dioxide fumigation is 20 times more effective at 96% r.h. than at 75% r.h.[21]. High humidities are to be avoided in plants such as roses, which are particularly susceptible to fungal diseases[22]. Changes in relative humidity can also create plant damage if the change occurs quickly. Soft-stemmed flowers such as gerberas and freesias when young and tender, or when growing under weak winter light, can suffer collapse of the stem tissue behind the head if the relative humidity falls quickly[23].

9.5 Conclusions

The importance of humidity in horticulture is growing. The use of dehumidifiers or ventilation control to avoid the high risk of plant disease at high relative humidities is now recognized. The concept that there may be an optimum relative humidity for growth which is different for each type of plant is now developing. Once the best growing conditions are realized then it is likely that the greenhouse controller will be a more direct analogue of transpiration rate, for example by using an artificial leaf rather than simple relative humidity control[24].

A simple guide to humidity and horticulture is presented in *Figure 9.10*.

References

1 Salisbury, F.B. and Ross, C. *Plant Physiology*. Wadsworth Publishing Co., California, USA, 1969
2 Clegg, C.J. and Cox, G. *Anatomy and Activities of Plants*. John Murray, London, 1978
3 O'Leary, J.W. and Knecht, G.N. 'The effect of relative humidity on growth, yield and water consumption of bean plants'. *J Am Hort Sci*, **96**(3), 263–265, 1971
4 Condar Manufacturing Co., USA, 1985
5 Scott-Russell, R. and Barber, D.A. 'The relation between salt uptake and the absorption of water by intact plants'. *Ann Rev Plant Physiol*, **11**, 127–140, 1960
6 O'Leary, J.W. 'The transporation stream and upward translocation of mineral ions'. *Ohio J Sci*, **65**, 357–362, 1970
7 Barletta, M. 'Four weeks of high r.h. affects ten trusses'. *The Grower*, May 28th, 1987
8 Winspear, K.W., Postlethwaite, J.D. and Cotton, R.F. 'The restriction of *Cladosporum fulvum* and *Botyritis cinerea* attacking glasshouse tomatoes, by automatic humidity control'. *Ann Appl Biol*, **65**, 75–83, 1970
9 Lipton, W.J. 'Growth of tomato plants and fruit production in high humidity and at high temperature'. *J Am Soc Hort Sci*, **95**(6), 674–680, 1970
10 Krizek, D.T., Bailey, W.A. and Klueter, H.H. 'Effect of relative humidity and type of container on the growth of F_1 hybrid annuals in controlled environments'. *Am J Bot*, **58**(6), 544–554, 1971
11 Hoffman, G.J., Rawlins, S.L., Garber, M.J. and Cullen, E.M. 'Water relations and growth of cotton as influenced by salinity and relative humidity'. *Agron J*, **63**, 822–826, 1971
12 Dalton, J. 'Experimental essays on the constitution of mixed gases'. *Memoirs and Proceedings of the Manchester Literary and Philosophical Society*, **5**, 535–602, 1802
13 Penman, H.L. 'Natural evaporation from open water, bare soil and grass'. *Proc R Meteorol Soc A*, **193**, 120–145, 1948
14 Monteith, J.L. 'Evaporation and environment', pp. 205–234 in *The State and Movement of Water in Living Organisms*. 19th Symp Soc Exp Biol, 1965
15 Rijtema, P.E. 'An analysis of actual evaporative transpiration'. Agric Res Report 659. Pudoc, Wageningen, The Netherlands, 1965

16 Cannon, J.N., Krantz, W.B., Kreith, F. and Naot, D. 'A study of transpiration from porous flat plates simulating plant leaves'. *Int J Heat Mass Transfer*, **22**, 469–483, 1979
17 Chau, K.V., Gaffney, J.J. and Romers, R.A. 'A mathematical model for the transpiration from fruits and vegetables'. *ASHRAE Trans*, **94**, Part I, 1988. Paper DA-88-20-1 RP-370
18 Rawson, H.M., Begg, J.E. and Woodward, R.G. 'The effect of atmospheric humidity on photosynthesis, transpiration and water use efficiency of leaves of several plant species'. *Planta (Berlin)*, **134**, 5–10, 1977
19 McLaughlin, S.B. and Taylor, G.E. 'Relative humidity: important modifier of pollutant uptake by plants'. *Science*, **211**, 167–169, 1981
20 Walls, I. *Modern Greenhouse Methods: Flowers and Plants*. F. Muller, London, 1982
21 Coney, H.M. and Uota, M. 'Effect of concentration, exposure time, temperature and relative humidity on the toxicity of sulphur dioxide to the spores of *Botrytis cinerea*'. *Phytopathology*, **51**, 815–819, 1961
22 Anon. *Roses under Glass*. Grower Guide No. 9. Grower Books, London, 1980
23 Anon. *Quality in Cut Flowers*. Grower Guide No. 11. Grower Books, London, 1980
24 Stanghellini, C. *Transpiration of Greenhouse Crops*. IMAG Instituut voor Mechanisatien, Wageningen, 1987

Further reading

Aubert, B. and Catsky, J. 'The onset of photosynthetic carbon dioxide influx in banana leaf segments as related to stomata diffusion resistance at different air humidities'. *Photosynthetica*, **4**, 254–256, 1970
Bailey, B.J. 'Limiting the relative humidity in insulated greenhouses at night'. *Acta Hort*, **148**, 1–11, 1983
Barrs, H.D. 'Controlled environmental studies of the effects of variable atmospheric water stress on photosynthesis, transpiration and water status of *Zea mays* L. and other species', pp. 249–258 in *Proc Uppsala Symp 1970* (ed. R.O. Slatyer). UNESCO, Paris, 1973
Bialoglowski, J. 'Effect of humidity on transpiration of rooted lemon cuttings under controlled conditions'. *Proc Am Soc Hort Sci*, **33**, 166–169, 1935
Bowen, L.S. 'The ratio of heat losses by conduction and evaporation from any water surface'. *Phys Rev*, **27**, 779–789, 1926
Breazeale, E.L. and McGeorge, W.T. 'Influence of relative humidity on root growth'. *Soil Sci*, **76**, 361–365, 1953
Brodie, H.J. 'Further observations on the mechanism of germination of conida of various species of powdery mildew at low humidity'. *Can J Res*, **23**, 198–206, 1945
Brouwer, R. and de Wit, C.T. 'A simulation model of plant growth with special attention to root growth and its consequences', in *Root Growth* (ed. W.J. Whittington). Butterworths, London, 1968
Demidenko, T.T. and Golle, V.P. 'Influence of relative air humidity on the yield and uptake of nutrient elements by the sunflower'. *CR Acad Sci URSS*, **25**, 328–332
Eberhardt, P.H. 'Influence de l'air sec at l'air humide sur la forme et sur la structure de vegetaux'. *Ann Sci Nat Bot Ser 8*, No. 18, 61–152, 1903
Ford, M.A. and Thorne, G.N. 'Effects of atmospheric humidity on plant growth'. *Ann Bot (Lond) NS*, **38**, 441–452, 1974
Gavande, S.A. and Taylor, S.A. 'Influence of soil, water potential and atmospheric evaporative demand on transpiration and the energy status of water in plants'. *Agronomy J*, **59**, 4–7, 1967
Grace, J. 'Wind damage to vegetation'. *Curr Adv Plant Sci*, **17**, 883–894, 1975
Hales, J.L. 'Review of calibration methods for humidity sensors', pp. 76–94 in *Proceedings of Conference Humidity Sensors and their Calibration* (ed. W.H. McGivern). NPL, Sept 1986
Hoffman, G.J. and Herkelrath, W.N. 'Infertility of cotton flowers at both high and low relative humidities'. *Crop Sci*, **10**, 721–723, 1970
Idso, S.B. 'Stomatal regulation of evaporation from well watered plant canopies: a new synthesis'. *Agric Meteorol*, **29**, 213–217, 1974
Kinbacher, E.J. 'Effect of relative humidity on the high temperature resistance of winter oats'. *Crop Sci*, **2**, 437–440, 1962
Lange, P.L., Losch, R., Schulze, E.D. and Kappen, L. 'Response of stomata to changes in humidity'. *Planta (Berlin)*, **100**, 76–86, 1971
Macara, T.J.R. 'The growth of mould on dried meat'. *J Soc Chem Ind*, **62**, 104–109, 1943

Matthews, R.B. and Saffell, R.A. 'Computer control of humidity in experimental glasshouses'. *J Agric Eng Res,* **33**, 213–221, 1986

Meurs, van W.T.M. and Gieling, T.H. 'A research strategy to solve air humidity problems in greenhouses caused by the use of energy saving measures'. *Acta Hort,* **106**, 77–83, 1980

Mitchell, J.W. 'Effect of atmospheric humidity on rate of carbon fixation by plants'. *Bot Gaz,* **98**, 87–104, 1936

Moldau, Kh. A. and Syber, A. Yu. 'Effect of air humidity on the conductivity of stomates and mesophyll of bean leaves at two values of soil moisture'. *Soviet Plant Physiol,* **21**, 663–668, 1974

Morris, L.G. and Winspear, K.W. 'The control of temperature and humidity in glasshouses by heating and ventilation'. *Proc Agric Eng Symp J Inst Agric Eng,* 1967

Nightingale, G.T. and Mitchell, J.W. 'Effect of humidity on metabolism in tomato and apple'. *Plant Physiol,* **9**, 217–236, 1934

Otto, H.W. and Daines, R.H. 'Plant injury by air pollutants: influence of humidity on stomatal apertures and plant response to ozone'. *Science,* **163**, 1209–1210, 1969

Peters, D.B. 'Growth and water absorption of corn as influenced by soil moisture tension, moisture content and relative humidity'. *Proc Soil Sci Soc Am,* **24**, 523–528, 1960

Pragnell, R.F. 'Review of humidity sensors', pp. 3–11 in *Proc of Conf NPL Humidity Sensors and their Calibration* (ed. W.H. McGivern). NPL Sept 1986

Sale, P.J.M. 'Growth and flowering of cacao under controlled atmospheric relative humidity'. *J Hort Sci,* **45**, 119–132, 1970

Sastry, S.K., Baird, C.D. and Buffington, D.E. 'Transpiration rates of certain fruits and vegetables'. *ASHRAE Trans,* **84**, Part I, 237–255, 1978

Schulze, E.D., Lange, O.L., Buschbom, U., Kappen, L. and Evenari, M. 'Stomatal responses to changes in humidity in plants growing in the desert'. *Planta (Berlin),* **108**, 259–270, 1972

Slatyer, R.O. and Bierhuizen, J.F. 'Transpiration from cotton leaves under a range of environmental conditions in relation to internal and external diffusive resistances'. *Aus J Biol Sci,* **17**, 115–130, 1964

Thut, H.F. 'Relative humidity variations affecting transpiration'. *Am J Bot,* **25**, 589–595, 1938

Waggoner, P.E. and Reifsnyder, W.E. 'Simulation of the temperature humidity and evaporation profiles in a leaf canopy'. *J Appl Meteorol,* **7**, 400–409, 1968

Woodward, R.G. and Begg, J.E. 'The effect of atmospheric humidity on the yield and quality of soya bean'. *Aust J Agric Res,* **27**, 501–508, 1976

Wright, N.C. 'The storage of artificially dried grass'. *J Agric Sci,* **31**, 194–201, 1940

Chapter 10

Agricultural factors

10.1 Seed storage

The ability of seeds to germinate and grow declines with time. This viability is strongly influenced by the temperature and moisture content of the seed. The moisture content is determined by the ambient relative humidity. This relationship is illustrated in *Figure 10.1*[1]. For any given relative humidity the moisture content depends upon the chemical nature of the seed. Starchy seeds such as oats, rice and barley are much more hygroscopic, because of the proteins, cellulose and starch within the seed, than the oily seeds of carrots, cabbages and tomatoes, which are rich in lipids. The adsorption isotherm is a physical characteristic of the seeds and is not related to the viability.

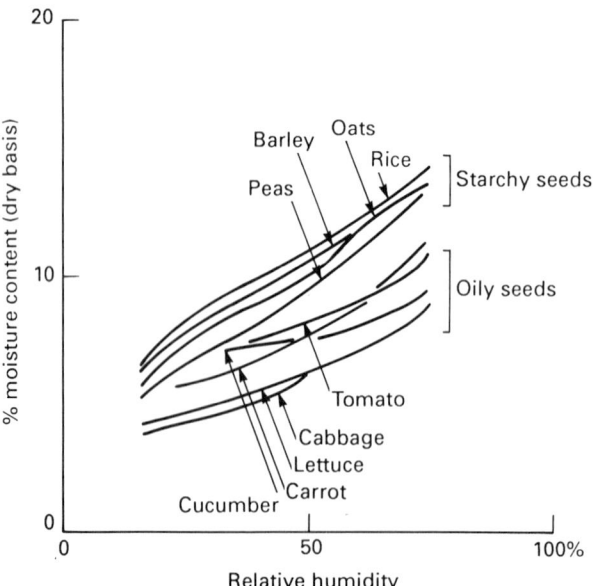

Figure 10.1 Adsorption isotherms for seeds at 25°C

Useful storage life increases as temperature and moisture content fall. For a given temperature, rule of thumb guidelines state that storage life is doubled for every 1% reduction in moisture content in the seed[2]. There are some exceptions to this, particularly for the seeds of hardwood trees, but these may be due to the speed of drying. Excessive drying speed splits and damages the protective shell of the seed. Within any given population of seeds there will be wide differences between individual seeds. This expresses itself as a survival curve for germination whereby the probability of germination can be linked to the storage time. This is illustrated for rice in *Figure 10.2*[3]. An alternative presentation style uses a probability scale which linearizes the probability function. This is termed probit analysis. The same data for rice in *Figure 10.3* are shown here as a straight line relationship.

While the probability analysis demonstrates the natural variability between individual seeds, there is no simple relationship for all seeds which shows the interrelationship between temperature and moisture content or

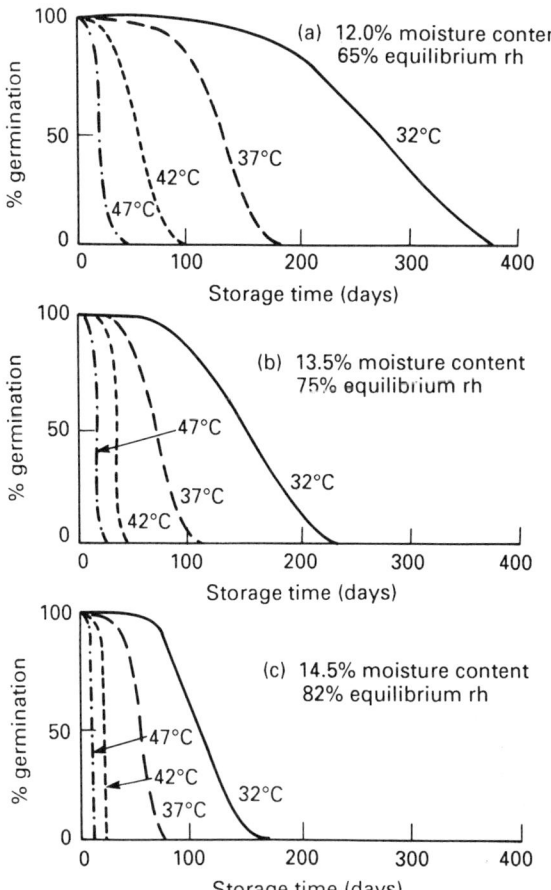

Figure 10.2 Survival curves for rice

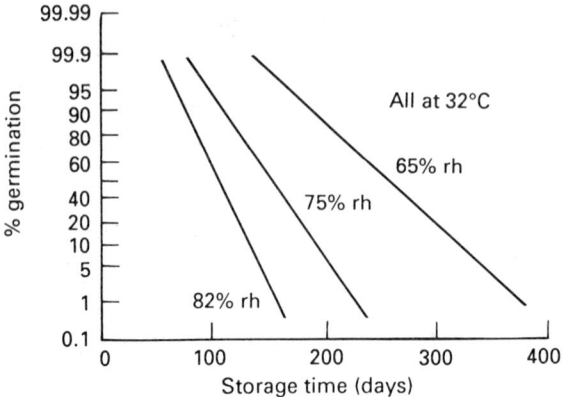

Figure 10.3 Survival curves for rice expressed on a probability scale

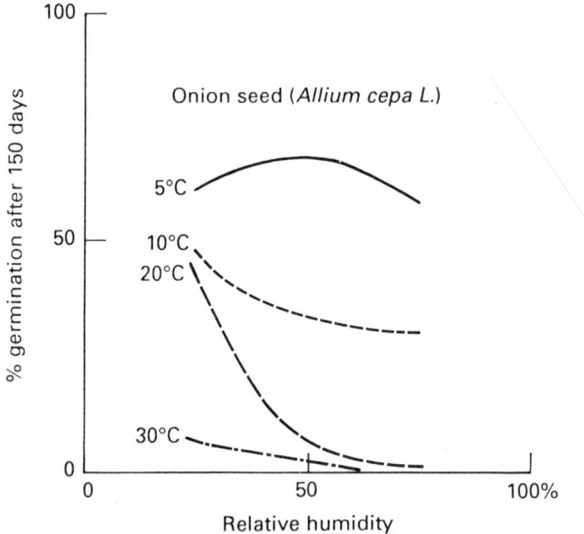

Figure 10.4 Germination of onion seeds after 150 days exposure to different temperature and humidity conditions

equilibrium relative humidity. A relatively delicate seed, onion, shows only a small change in germination with a wide change in ambient relative humidity, at low temperatures around 5°C. However, at 20°C the influence of relative humidity is dramatic. This phenomenon is illustrated in *Figure 10.4*[4]. The result of a separate study on onion seed which shows the variation in germination over time is given in *Figure 10.5*[5].

A more resilient seed such as tomato is much less sensitive to both temperature and humidity at temperatures between 5 and 20°C. However, at 30°C, viability declines rapidly at ambient relative humidities above 60%. This is illustrated in *Figure 10.6*. Lettuce seed, which is between tomato and onion in robustness, has a survival curve which is illustrated in

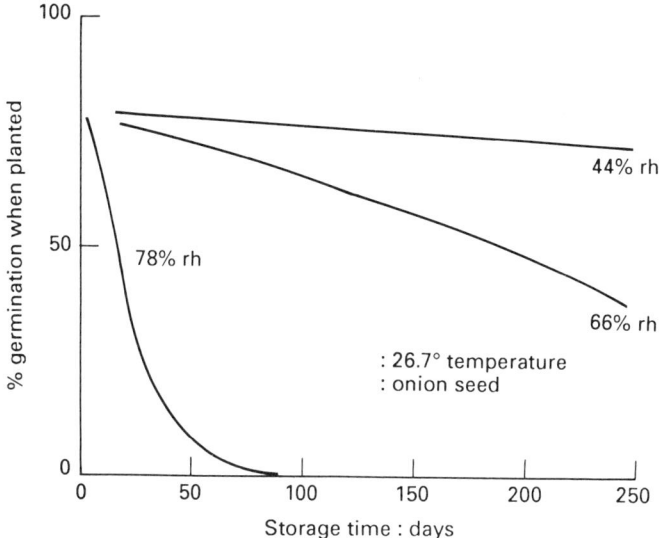

Figure 10.5 The influence of relative humidity on the germination of onion seeds after different periods of storage at 26.7°C

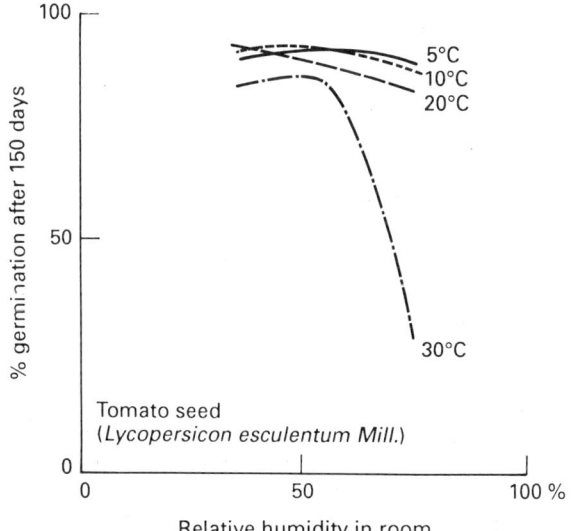

Figure 10.6 Germination of tomato seeds after 150 days exposure to different temperature and humidity conditions

Figure 10.7. Above 10°C the viability declines rapidly above an ambient relative humidity of 60%.

At relative humidities above 70% mould begins to grow. A study on one of the widespread fungal genuses, *Aspergillus*, showed that the presence of mould could lead to serious deterioration in viability. Peas were used as the seed and were stored at 30°C and 85% r.h. These were inoculated with one

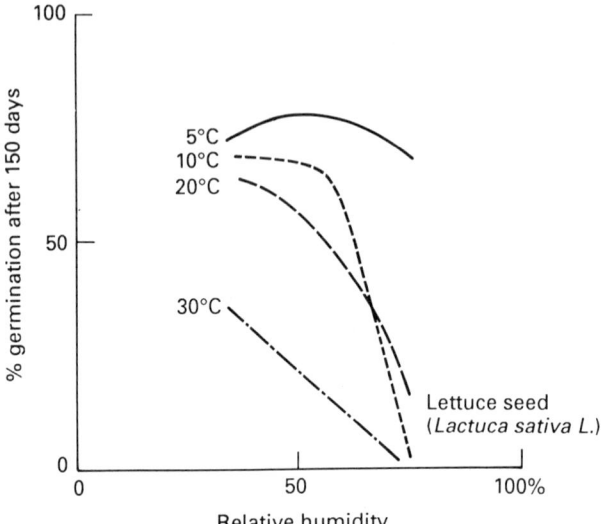

Figure 10.7 Germination of lettuce seed after 150 days exposure to different temperature and humidity conditions

of several *Aspergillus* species. Two important facts emerged. The first was that the moulds were toxic and that the most effective toxin was produced by *Aspergillus flavus*. These effects are demonstrated in *Figures 10.8* and *10.9*. The second factor was that the loss in viability was not directly linked to the degree of colonization. Mould which grew at 30°C produced a bigger reduction in viability than those at 20°C or 10°C for similar degrees of colonization[6]. More recent studies have shown *Aspergillus* to be one of the moulds which readily generate toxins[7]. Preliminary results suggest that the toxin production is highest when the mould is growing quickly. The toxins do not appear to be associated with the slowest growing conditions. Toxin production is believed to stop at ambient relative humidities below 83% while the mould can grow, albeit slowly, down to conditions where the ambient is 80% r.h.

While, in general, dryness helps to enhance viability, there is a limit of dryness at which desiccation injuries occur. This occurs around 2% moisture content, which for wheat would be an ambient relative humidity of a few per cent. This would be impractical for anything other than laboratory studies. Care is also needed in the speed of drying. The seed temperature should not exceed 45°C and the drying rate has to match the type of seed. Thicker seeds and seeds with tough outer cases require slow drying to avoid splitting. Beans, corn and peas need to be dried slowly. The smaller or more permeable seeds can dry quickly. This would include the seeds of grasses, rape and sugar beet.

Care is also needed if storage temperatures below freezing are encountered. Seeds can be damaged by frost. The damage is more severe if the seed has a high moisture content.

A good overall storage condition is 10°C, 40% r.h.

The storage requirements for seeds do not apply to the roots of

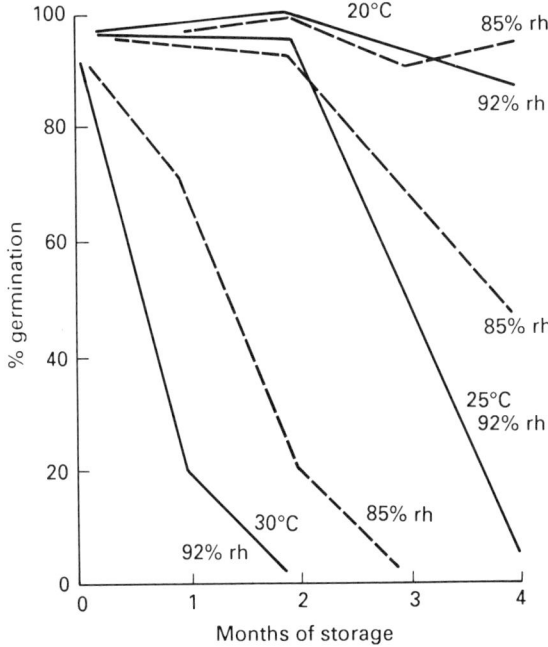

Figure 10.8 The effect of *Aspergillus flavus* mould on the germination of peas

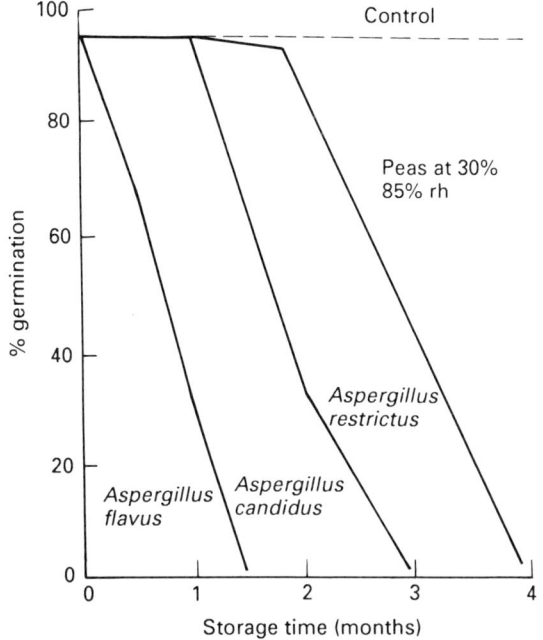

Figure 10.9 Moulds, particularly *Aspergillus flavus*, affect seed viability

frost-sensitive plants such as dahlias[8]. Such roots require careful conditioning and low temperatures. Ambient temperature determines when the root will start to sprout. Successful storage has to be between 4 and 7°C to inhibit sprouting. Once the storage temperature is correct then ambient relative humidity becomes important but in the opposite sense to that for seeds. Moist conditions are required for roots. When stored below 50% r.h. the roots will continue to lose moisture and the weight loss is over half after six months. Dahlias will not recover from such intense shrivelling. At 90% r.h. the roots will develop small white feeding roots which are needed during the growth phase of the plant. Ideal long-term storage conditions are 7°C, 80–85% r.h. This is illustrated in *Figure 10.10*.

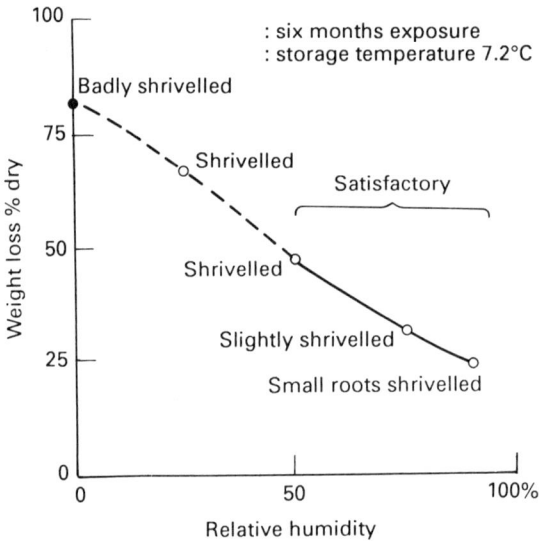

Figure 10.10 Weight loss of dahlia roots (Jersey Beauty) after six months exposure at 7.2°C and for a range of relative humidities

10.2 Farm animals

Farm animals have only recently been farmed indoors and knowledge of the conditions under which they thrive is developing rapidly. In general, animals thrive best when under little or no physiological stress. Temperature is usually the critical factor and humidity plays a secondary role, and even then only when the animal is overheating.

The relative importance of the evaporative component of heat loss is different for the different animals. Cows have the largest component, then pigs, sheep and poultry. This is illustrated in *Figure 10.11*.

Milk production progressively falls when ambient temperatures go below 5°C. Production is unaffected by the various meteorological factors when the temperature is between approximately 5 and 25°C. The upper temperature varies with the type of cow. Holsteins are much less heat

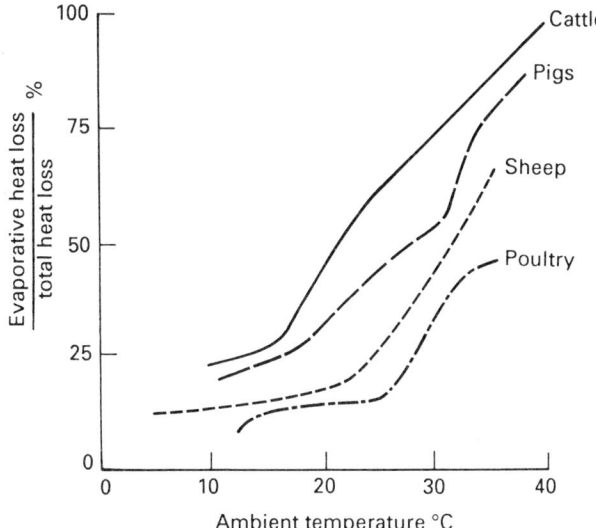

Figure 10.11 Evaporative loss is important at high temperatures

tolerant than Brown Swiss or Brahman cattle. Milk yield progressively declines for higher temperatures. The rate of decline is increased by increasing humidity and increasing sunshine. This is illustrated in *Figure 10.11*[10]. Laboratory experiments on Holstein cows show the effect of ambient relative humidity on milk yield. This is illustrated in *Figure 10.13*[9].

Poultry are quite tolerant to warm conditions but humidity strongly influences the amount of dust generated within the space. When the relative humidity was raised from 40% r.h. to 70% r.h. the amount of dust generated by each bird more than halved. This has important implications for the health of the hen and for the cost of air filtration. The age of the litter on the floor had an even more dramatic effect. Old litter, containing

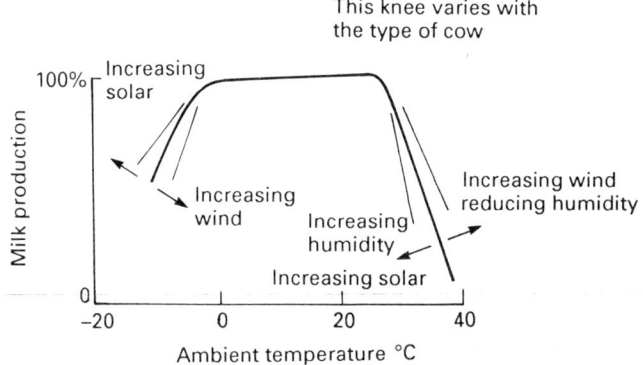

Figure 10.12 Milk production is influenced by the environment

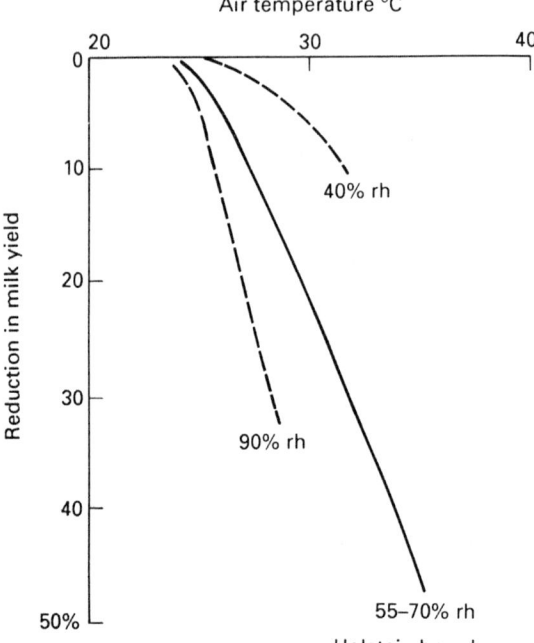

Figure 10.13 Milk production at high temperatures is influenced by humidity

droppings, food remains and skin scales, was much more prone to generate high dust burdens. The effect of increasing the ambient relative humidity was equally effective in suppressing this dust (*Figure 10.14*).

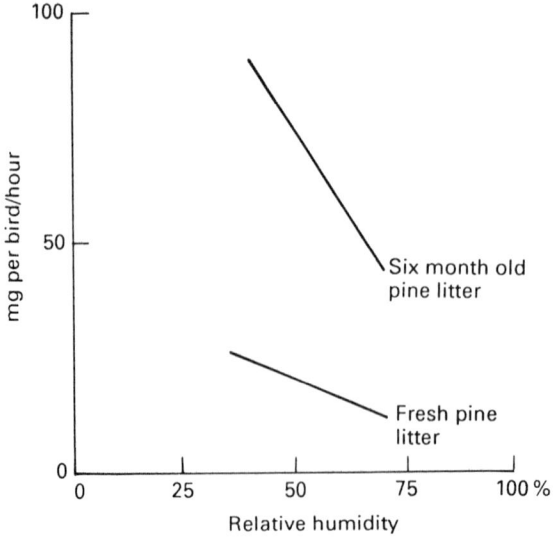

Figure 10.14 Dust generation in hen houses varies with humidity

References

1 Howe, R.W. Chapter 4 in *Seed Biology* (ed. T.T. Kozlowski). Academic Press, New York, 1972
2 Harrington, F.J.F. 'Thumb rules for drying seed'. *Crops and Soils,* **13**, 16–17, 1970
3 Roberts, E.H. (ed.). *Viability of Seeds.* Chapman & Hall Ltd, London, 1972
4 Barton, L.V. 'Relation of certain air temperatures and humidities to viability of seeds'. *Boyce Thompson Inst Contrib,* **12**, 85–102, 1941
5 Boswell, V.R., Toole, E.H., Toole, V.K. and Fisher, D.F. 'A study of rapid deterioration of vegetable seeds and methods for its prevention'. *US Dept Agric Tech Bull,,* 708, 1940
6 Fields, R.W. and King, T.H. 'Influence of storage fungi on deterioration of stored pea seed'. *Phytopathology,* **52**, 336–339, 1962
7 Troller, J.A. and Christian, J.H.B. *Water Activity and Food.* Academic Press, New York, 1978
8 Allen, R.C. 'Temperature and humidity requirements for the storage of Dahlia roots'. *Am Soc Hort Sci Proc,* **35**, 770–773, 1938
9 Yeck, R.G. and Stewart, R.E. 'A ten year summary of the psychroenergetic laboratory dairy cattle research at the University of Missouri'. *Trans Am Soc Agric Eng,* **2**, 71–77, 1959
10 Johnson, H. D. 'The response of animals to heat'. *Meteorological Monographs,* **6**, 110–121, 1965

Further reading

Barton, L.V. 'The storage of citrus seeds'. *Contr Boyce Thompson Inst,* **13**, 47–55, 1943
Bond, T.E., Kelly, C.F. and Heitman, H. 'Hog house air conditioning and ventilation data'. *Trans Am Soc Agric Eng,* **2**, 1–4, 1959
Brett, C.C. 'The influence of storage conditions upon the longevity of seeds with special reference to root and vegetable crops'. *Rep 13th Int Hort Congr 1952,* 1016–1018, 1953
Brody, S. 'Do cow barns need air conditioning?' *Refrig Eng,* 39–45, April 1956
Burges, H.D., Edwards, D.M., Burrell, N.J. and Cammell, M.E. 'Effects of storage temperature and moisture content on the germinative energy of malting barley, with particular reference to high temperature'. *J Sci Food Agric,* **15**, 32–50, 1964
Christensen, C.M. 'Microflora and seed deterioration', pp. 173–190 in *Viability of Seeds* (ed. E.H. Roberts). Chapman & Hill, London, 1972
Christensen, C.M. and Lopez, F. 'Pathology of stored seeds'. *Proc Int Seed Test Assoc,* **28**, 701–711, 1963
Cooke, F.M. 'Humidity control of seed storage areas'. *Miss Seedsmen Assoc: Short Course Seedsmen Proc,* 29–37, 1966
Crocker, W. 'The life span of seeds'. *Bot Rev,* **4**, 235–274, 1948
Crocker, W. and Barton, L. *Physiology of Seeds.* Chronica Botanica Co., USA, 1950
Darsie, M.L., Elliott, C. and Peirce, G.H. 'A study of the germinating power of seeds'. *Gazette,* **58**(2), 101–136, 1914
Gane, R. 'The water content of the seeds of peas, soy beans, linseed, grass, onion and carrot as a function of temperature and humidity of the atmosphere'. *J Agric Sci,* **38**, 81–89, 1948
Groves, J.F. 'Temperature and life duration of seeds'. *Bot Gazette,* **63**, 169–189, 1917
Grub, W., Rollo, C.A. and Howes, J.R. 'Dust problems in poultry environments'. *Trans Am Soc Agric Eng,* **8**(5), 338–339, 352, 1965
Harrington, J.F. 'Seed storage and longevity', Chapter 3, pp. 145–245 in *Seed Biology* (ed. T.T. Kozlowski). Academic Press, New York, 1972
Heitman, H., Kelly, C.F. and Hughes, E.H. 'Californian psychrometric chamber for livestock environmental studies'. *J Animal Sci,* **8**, 450–463, 1949
James, E. 'Preservation of seed stock'. *Adv Agron,* **19**, 87–106, 1967
James, E., Bass, L.N. and Clark, D.C. 'Varietal differences in longevity of vegetable seeds and their response to various storage conditions'. *Am Soc Hort Sci,* **91**, 521–528, 1967
Jones, J.W. 'Germination of rice seed as affected by temperature, fungicides and age'. *J Am Soc Agron,* **18**, 576–592, 1926
Justice, O.L. and Bass, L.N. *Principles and Practices of Seed Storage.* Castle House Publications, London, 1979
Koon, J., Howes, J.R., Grub, W. and Rollo, C.A. 'Poultry dust: origin and composition'. *Agric Eng,* **44**(11), 608–609, November 1963

Kreyger, J. 'Recherches sur la conservation des orges de brasserie'. *Le Petit Journal du Brasseur*, **67**, 7–10, 1959

Larmour, R.K., Sallans, H.R. and Craig, B.M. 'Hygroscopic equilibrium of sunflower seed, flax seed and soy beans'. *Can J Res*, **22**(1), 1–8, 1944

Lopez, L.C. and Christensen, C.M. 'Factors influencing the invasion of *Sorghum* seed by storage fungi'. *Plant Disease Reporter*, **47**, 597–601, 1963

Malone, J.P. and Muskett, A.E. 'Seed borne fungi'. *Proc Int Seed Test Assoc*, **29**, 179–384, 1964

Mitchell, F.S. and Cadwell, F.Y.K. 'Influence of variations in harvesting and initial storage on wheat kept for several years'. *J Agric Eng Res*, **7**, 27–41, 1962

Mudalier, C.R. and Sundararaj, D.D. 'Dormancy and germination of a few crop seeds'. *Madras Agric J*, **41**, 111, 1954

Nutile, G.E. 'Effect of desiccation on viability of seeds'. *Crop Sci*, **4**, 325–328, 1964

Owen, E.B. 'The storage of seeds for the maintenance of viability'. *Commonwealth Bur of Pasture and Field Crops Bull*, No. 43, 1956

Parkin, E.A. 'Stored product entomology'. *Ann Rev Entomol*, **1**, 223–240, 1956

Qasem, S.A. and Christensen, C.M. 'Influence of various factors on the deterioration of stored corn by fungi'. *Phytopathology*, **50**, 703–709, 1960

Ragsdale, A.C., Thompson, H.J., Worstell, D.M. and Brody, S. 'The effect of humidity on milk production and composition of feed and water consumption and body weight in cattle'. *Research Bulletin No. 521, Series 21*. Missouri Agricultural Experiment Station, 1953

Roberts, E.H. 'The viability of cereal seed in relation to temperature and moisture'. *Ann Bot*, **24**(93), 12–31, 1960

Roberts, E.H. 'The viability of rice seed in relation to temperature, moisture content and gaseous environment'. *Ann Bot*, **25**(99), 381–390, 1961

Roberts, E.H. 'Viability of cereal seed for brief and extended periods'. *Ann Bot*, **25**(99), 373–380, 1961

Roberts, E.H. and Abdalla, F.H. 'The influence of temperature, moisture and oxygen on period of seed viability in barley, broad beans and peas'. *Ann Bot*, **32**, 97–117, 1968

Robertson, D.W., Lute, A.M. and Gardner, R. 'Effect of relative humidity on viability, moisture content and respiration of wheat, oats and barley seed in storage'. *J Agric Res*, **59**(4), 281–291, 1939

Sahadwan, P.C. 'Studies on the problem of loss of viability of rice seeds in storage'. *Madras Agric J*, **40**, 133–143, 1953

Sayre, J.D. 'Storage tests with seed corn'. *Farm and Home Research*, **32**, 149–154, 1947

Toole, E.H., Toole, V.K. and Gorman, E.A. 'Vegetable seed storage as affected by temperature and relative humidity'. *US Department of Agriculture Technical Bulletin No. 972*, 1948

Touzard, J. 'Influences de diverses conditions constanter de temperature et d'humidité sur la longévité des graines de quelques espèces cultivées. Advances in horticultural science and their applications'. *Proc 15th Int Hort Congress Nice*, **1**, 339–347. Pergamon, Oxford, 1961

US Department of Agriculture. 'Poultry respiration colorimetric studies of laying hens'. *Agric Research Service Report*, 42–43, June 1961

Vegis, A. 'Climatic control of germination, bud break and dormancy', pp. 265–287 in *Environmental Control of Plant Growth* (ed. L.T. Evans). Academic Press, New York, 1963

Virgin, W.J. 'Low germination of peas associated with the presence of bacteria in seed'. *Phytopathology*, **30**, 790–791, 1940

Webster, L.V. and Dexter, S.T. 'Effects of physiological quality of seeds on total germination, rapidity of germination and seedling vigour'. *Agron J*, **53**, 297–299, 1961

Chapter 11

Moisture control to avoid grain and flour spoilage

11.1 Introduction

Food such as the annual grain harvest is stored in its natural form in large amounts for long periods. Grain is drawn from this store and milled to provide flour which in turn is stored, but for shorter periods and usually in smaller quantities.

A whole range of insects, beetles, cockroaches, weevils and mites are likely to be present in very small quantities in the grain store, and the storage process should ensure that these insects or their eggs do not multiply and create an infestation. Such an infestation can make the product look unattractive and smell unpleasant. It can also reduce the nutritional value and introduce disease vectors, particularly for the people who work with the material.

The conditions under which an infestation is likely to occur are principally influenced by temperature and then by moisture. There is a relationship between the moisture content of the food and the ambient relative humidity of the surrounding atmosphere. An example for wheat is shown in *Figure 11.1*. This is called an adsorption isotherm[1] and it will vary a little with temperature. We can therefore define moisture criteria in terms of either moisture content of the material or ambient equilibrium relative humidity. Wherever possible, data in this section are given in terms of relative humidity and temperature rather than moisture content of the food.

Insects live at a temperature very close to that of their surroundings. Their activity level in terms of both rate of development and of reproduction increases with rising temperature up to a temperature of around 35°C. Few eggs are laid at temperatures above 35°C and adult life is usually limited. At lower temperatures the activity rate slows down; few eggs are laid below 15°C, and below 5–10°C dormancy sets in. Below freezing point the insects are usually killed after a few weeks. This temperature regime is illustrated in general in *Figure 11.2*[2].

Moisture becomes critical at those temperatures at which reproduction can occur, i.e. typically between 15 and 35°C. Each type of insect has a different requirement, which we will now examine[3–5]. We will consider

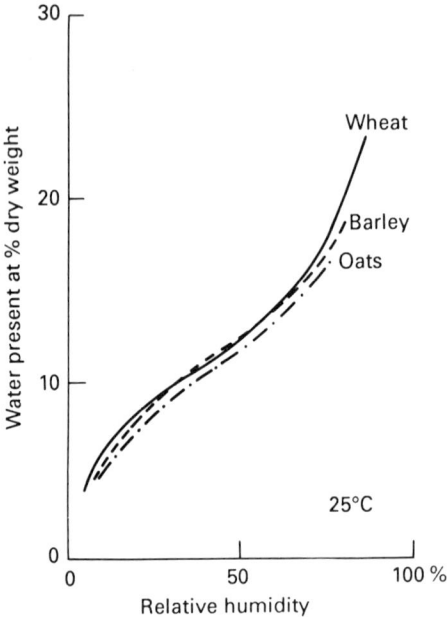

Figure 11.1 The relationship between moisture content of wheat and ambient relative humidity

them generally first and then in terms of decreasing size: cockroaches, beetles, weevils and mites.

11.2 Water balance in insects

Water is essential for the continued life of an insect and represents approximately 50% of its weight. If this water is progressively lost then the insect becomes less active and will eventually die through desiccation.

The two principal sources of water are:

(1) *Food intake.* Only a small number of food pests drink. These include some moths and cockroaches. The majority of insects ingest moisture with their food. It is quite common for dry-looking material, such as grain or flour, to be between 5 and 15% water by weight.

(2) *Metabolism.* Complete oxidation of both carbohydrates and fats generates carbon dioxide and water. For a grain-eating insect, a typical oxidation process would be [6]:

For a carbohydrate (glucose)

glucose + oxygen   = carbon dioxide + water + energy
$C_6H_{12}O_6 + 6O_2$ = $6CO_2$ + $6H_2O$
180 g + 134.4 litres = 134.4 litres   + 108 g + 677 k cals

i.e. 108 g of water are created for every 180 g of carbohydrate oxidized.

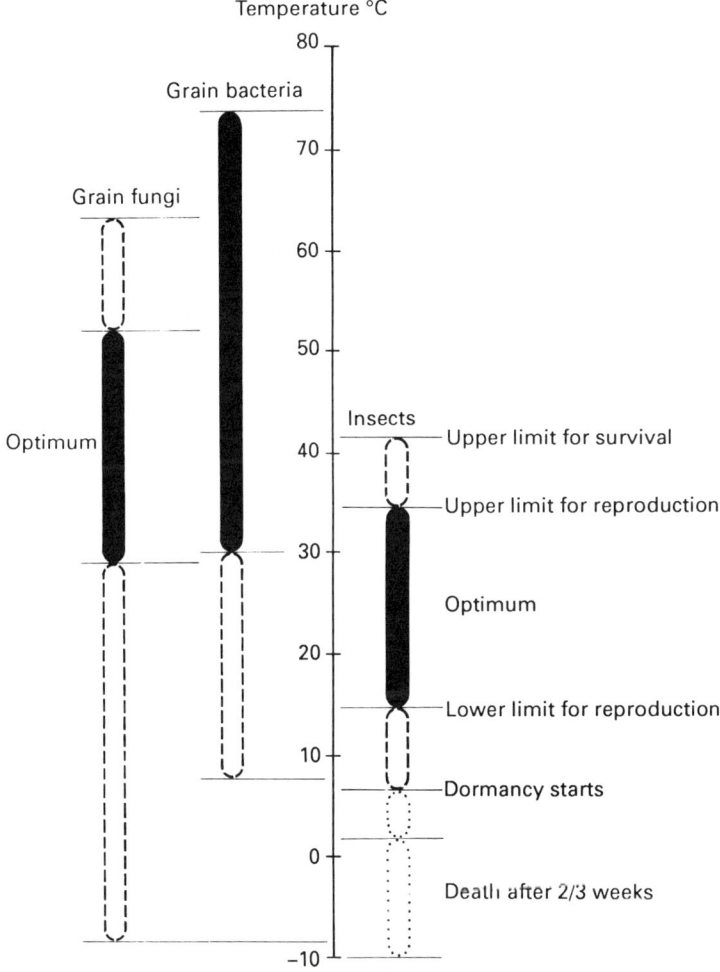

Figure 11.2 Approximate temperature limits for grain pests

For a fat (tripalmitin)

tripalmitin + oxygen = carbon dioxide + water + energy

$(C_{15}H_{31}COO)_3C_2H_5$ + $72.5O_2$ = $51\,CO_2$ + $49H_2O$

807 g + 1642 litres = 1142 litres + 883 g +

7617 k cal

i.e. 883 g of water are generated for every 807 g of fat oxidized.

This illustrates that the quantity of moisture generated within the insect itself is almost equal to the weight of the food it eats. This is supplemented by the moisture within the food itself.

The two routes for moisture losses are:

(1) *Evaporation.* Most insects have a respiration system based on a large number of air tubes which penetrate deeply into the body. The oxygen in

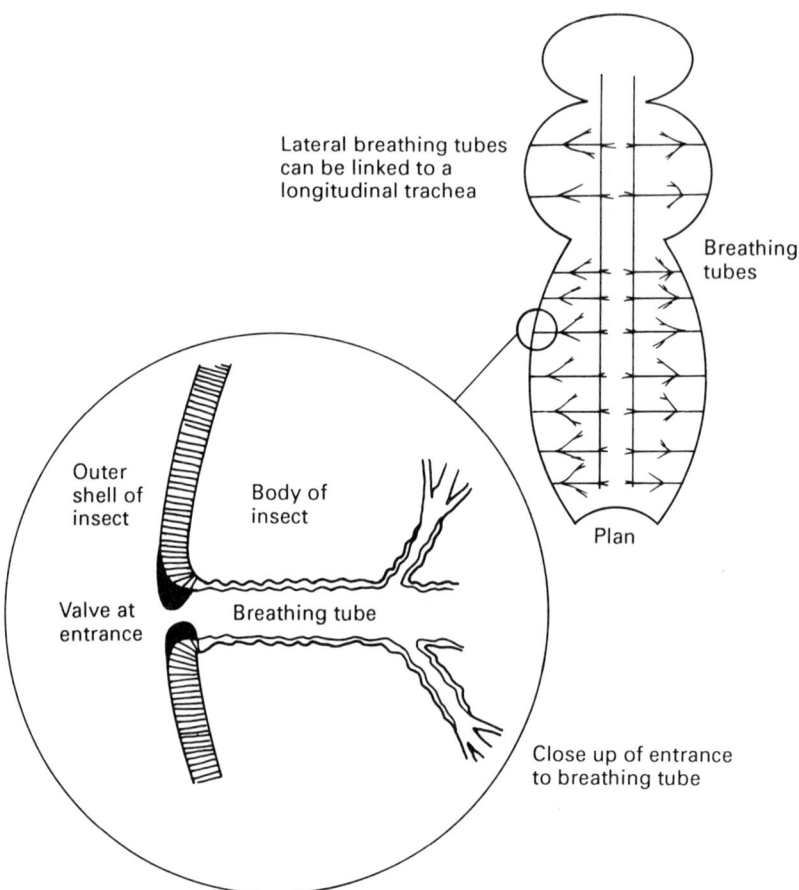

Figure 11.3 Most insects breathe through tubes which penetrate into the body of the insect

the air diffuses into these tubes to the blood and carbon dioxide is released through the same tubes (*Figure 11.3*)[7]. The mechanism is normally diffusion-controlled, with a special valve in the shell of the insect which regulates the degree to which the entrance of the tube is open. When the insect is highly active or in very hot environments, this diffusion of air into and out of the spiracles can be enhanced by a pumping action of the body shell itself. When the insect is dormant or in very dry atmospheres, the valves close and effectively stop evaporation[8].

(2) *Faecal pellets.* Small amounts of moisture are lost through the ejection of faecal pellets by the insect. This loss is normally insignificant in comparison with evaporation.

The major moisture loss is therefore evaporation, which in turn is related to the water vapour pressure deficit in the atmosphere. This deficit is the vapour pressure difference between the saturated water vapour in the breathing tubes, which are at the temperature of the insect, and the vapour pressure in the ambient air (*Figure 11.4*). High deficits can only be incurred at low ambient relative humidities and high temperatures.

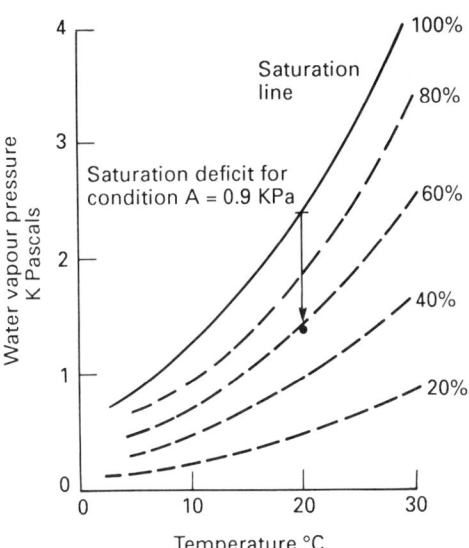

Figure 11.4 Psychrometric chart showing saturation water vapour pressure deficit for condition A

11.3 Cockroaches

Cockroaches are a kitchen pest and because of their desire for a high temperature (21–33°C) and their long development time of three months at 21°C, they tend to develop in continuously heated kitchens serving food all round the day, as in a hospital.

The German cockroach is the smallest and most common type (*Figure 11.5*). It is omnivorous, gregarious and normally only active at night. Measurements of moisture loss show a clear relation with the saturation water vapour deficit (*Figure 11.6*) for temperatures up to 30°C. At higher temperatures the cockroach moves its body to mechanically ventilate the breathing tubes to keep cool.

They are not normally sensitive to humidity because they can drink water and recover 30% body weight loss very quickly. They can survive up to 60% weight loss by desiccation[9, 10].

They destroy food stocks by fouling and tainting them with an unpleasant smell. They are particularly mobile, and while their wings will not permit flight, they are used for gliding quickly from high points to the floor. This mobility means that contamination can be spread widely.

11.4 Beetles

A wide variety of beetles attack stored grain and flour. These include the saw-toothed grain beetle, the flat grain beetle, the lesser grain borer, the red flour beetle and the confused flour beetle. Their population is

Name	Length	Illustration	Diet
German cockroach (*Blattella orientalis*)	15 mm		Omnivorous requires warmth
Saw tooth grain beetle (*Oryznephilus surinamensis*)	3 mm		– grain stores, dried food, nuts – will not breed below 18°C
Flat grain beetle (*Cryptolestes ferrugineus*)	2 mm		– grain stores need high temp. and humidity
Flour beetle (*Tribolium confusum*)	4 mm	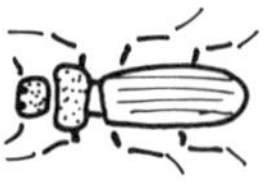	– flour, cereals, dried fruit, chocolate – will not breed below 18°C
Grain weevil (*Sitophilus granarius*)	4 mm	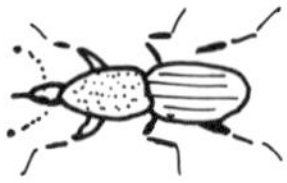	– stored grain
Flour mite (*Acarus siro*)	1 mm		– flour seed and corn – breeds down to 4°C – requires high humidity 65% rh

Figure 11.5 Common grain pests

determined by the climate in which the grain was grown and by the temperature and humidity of the store. Since most of the pests are of subtropical origin, they become inactive at storage temperatures below 20°C, and do little damage. Above this temperature their reproduction rate is strongly influenced by the ambient relative humidity, which determines the moisture content of the grain.

The flour beetles can reproduce in flour or grain dust from which practically all the moisture has been removed. They will also attack a variety of other dried foods such as beans, fruit, spices and chocolate. The most common type is the confused flour beetle. Flour infested by the larvae becomes grey and has an enhanced tendency to go mouldy. The

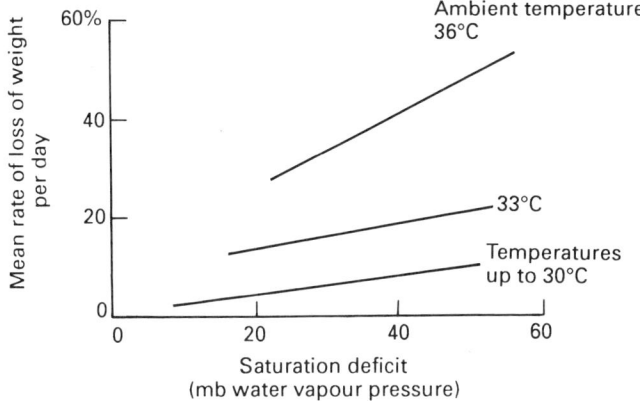

Figure 11.6 Moisture loss from the cockroach

beetles also contaminate the flour with their faecal pellets, which contain uric acid. They also secrete a pungent, irritating liquid through special odoriferous glands. The preferred temperature is 30°C and the confused flour beetle will not develop or breed at temperatures below 18°C.

The flour beetles will attack the whole kernel grain but only if the moisture content is high. At 27°C, experiments on whole grain showed that the confused flour beetle required a relative humidity of 50% before it could survive (12% moisture content in the wheat). The survival rate fell rapidly with time at lower relative humidities and all had died after 100 days at 20% r.h. (8% moisture content in the wheat) (*Figure 11.7*). However, when powdered grain dust was added to clean wheat, the flour beetles survived even at 20% r.h.

The rust red flour beetle is similar to the confused flour beetle but is a little smaller and darker and requires warmer conditions. Experiments on 25 pairs of this beetle in wheat showed how the ambient temperature and

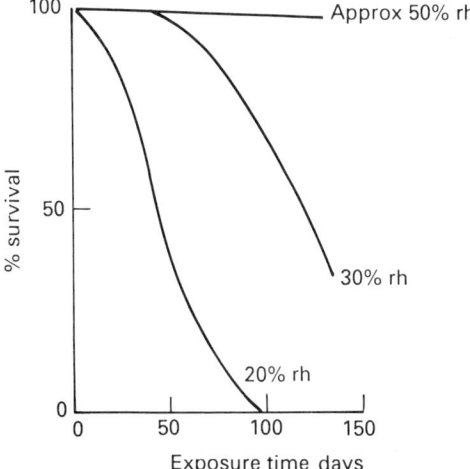

Figure 11.7 Survival of the flour beetle (*T. confusum*) in clean wheat

humidity affected the rate of multiplication. Below 25°C and 30% r.h. the beetles did not significantly multiply. At temperatures up to 27°C there was no multiplication until the relative humidity exceeded 60% (13% moisture content of the grain) (*Figure 11.8*)[11].

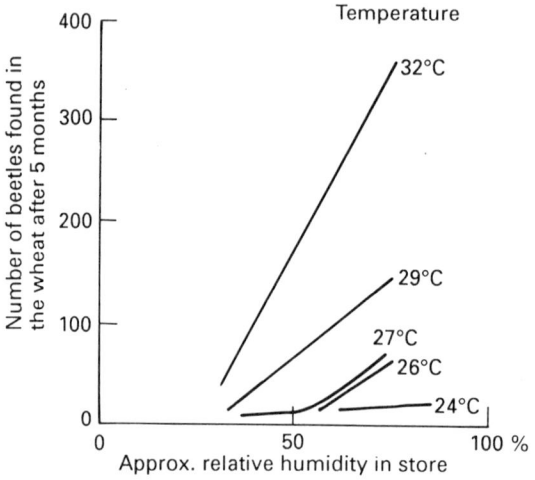

Figure 11.8 Abundance of beetles in wheat stored at different temperatures and humidities

Investigations[12] of the effect of temperature and humidity on the number of eggs laid by the female red flour beetle showed a significant increase in eggs laid by the beetles in the 70% r.h. atmosphere, compared with those at 30%. A special check test was made at 2% r.h. and this was very similar to the value at 30% r.h. These results are illustrated in *Figure 11.9*. The proportion of eggs which hatched averaged 90% and was unaffected by ambient humidity.

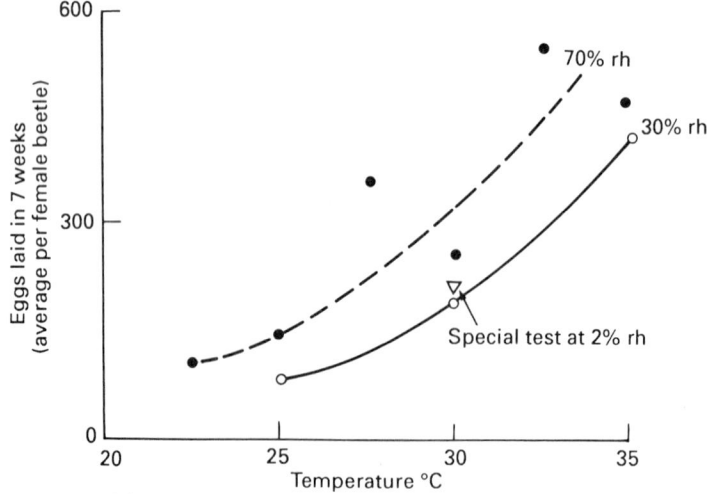

Figure 11.9 Effects of temperature and humidity on the number of eggs produced by the female red flour beetle during the first seven weeks of laying

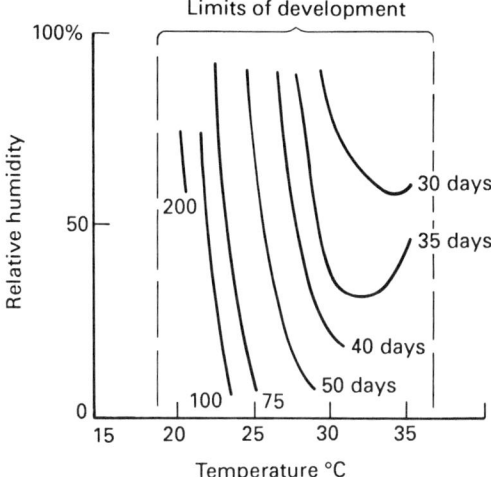

Figure 11.10 Development time of the adult flour beetle (*Tribolium madens*) from the time of egg laying

An illustration of the development time for the formation of the adult beetle from the egg is illustrated in *Figure 11.10* for *Tribolium madens*. The more common confused flour beetle has a rather shorter development time at lower temperatures.

11.5 Weevils

The granary weevil is the most common pest in stored grain. Its development route makes infestation difficult to detect in its early stages. The female gnaws a hole in an individual grain, lays an egg in it and seals it with a secretion which is similar in colour to the grain kernel. The larva cannot leave the grain but eats over half of the inside of it before it pupates, still inside the husk, and emerges a week later as a fully developed weevil. The granary weevil cannot fly but can wander long distances[13].

The rice weevil is slightly smaller in size and is a serious pest in tropical and subtropical countries. It can fly. It favours warm conditions and is not hardy in cold weather.

Experiments on the influence of temperature on the development of the weevil population in a bed of grain to which 50 pairs of weevils had been introduced are shown in *Figure 11.11*. Two experiments were performed, one on granary weevils and the other on rice weevils. Both granary weevils and rice weevils were most prolific at 27°C. At 20°C the rate of breeding of the rice weevil was less than one-quarter that of the granary weevil. When these experiments were continued over a range of relative humidities at the optimum temperature, both types of weevil, granary and rice, behaved similarly. Multiplication only occurred at relative humidities above 40–50% (*Figure 11.12*). Survival time fell rapidly at lower relative humidities (*Figure 11.13*)[4].

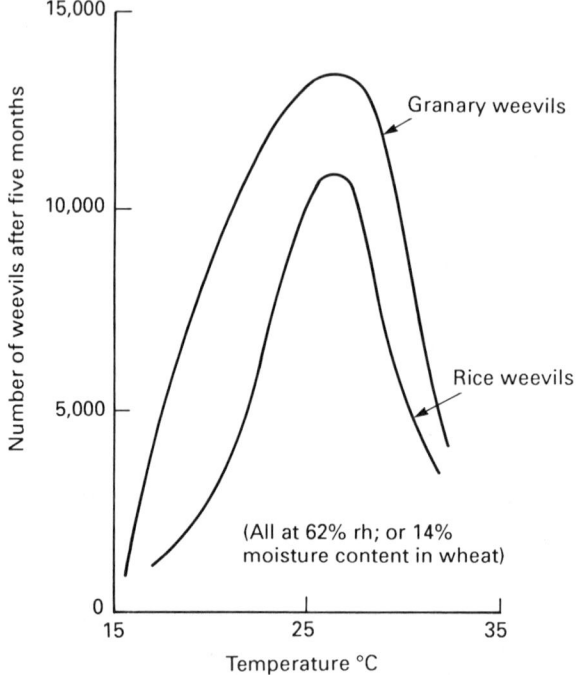

Figure 11.11 The reproduction of rice and granary weevils in clean wheat

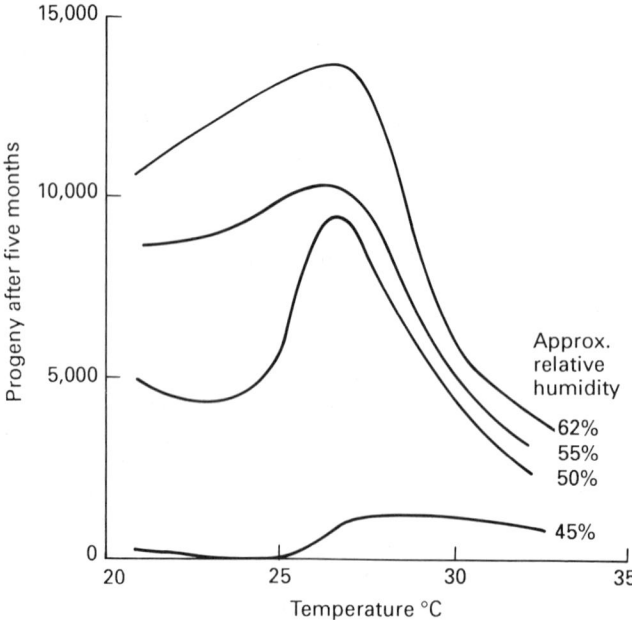

Figure 11.12 Reproduction of rice weevils from 100 weevils introduced into clean wheat

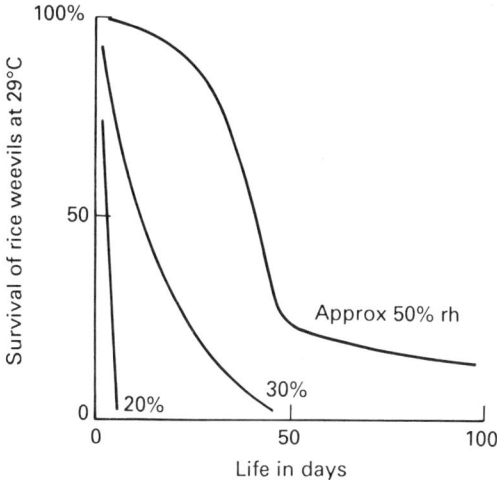

Figure 11.13 Survival of adult rice weevils in wheat

11.6 Mites

Mites are amongst the tiniest of insects (*Figure 11.5*). They often lack the breathing tube spiracles of the larger insects and their gas exchange mechanism is directly through the thin cuticle surrounding their bodies. This makes them particularly susceptible to desiccation in dry conditions, because the insect cannot control its moisture loss.

The flour mite is the most common mite in food stores. The female is prolific in egg laying and can lay 500 eggs in her lifetime. After hatching, the life cycle comprises a larval stage, followed by two nymphal stages, where the small mites start to look like the adult, and finally the adult. At 25°C the development time is three weeks. However, the mites can complete their development even at temperatures down to 4°C. Fortunately, they do require moist conditions and will not develop below 65% r.h. They are a persistent pest because during the second nymphal stage they can enter a dormant phase which is very resistant to desiccation and can revive and develop when damper conditions return[14-17]. Limits and rates of reproduction for different temperatures and humidities are illustrated in *Figure 11.14*[18].

Mite-infested food has a sweet, sickly smell and an undesirable taste, which discourages its use for human consumption. If infested flour is used for bread-making it has a sour taste and poor colour, and may not rise properly. Mites also induce dermatitis in people who regularly handle the infested products and can be responsible for some allergies[19, 19a].

11.7 Microflora

Moulds and bacterial rot can decrease the viability of grain, cause unsightly discoloration, introduce mustiness and unpalatability, create hot spots in

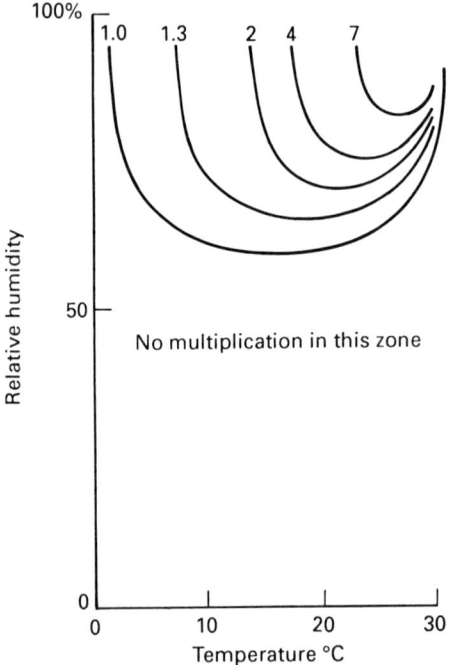

Figure 11.14 Multiplication of the flour mite (*Acarus siro*) after one week

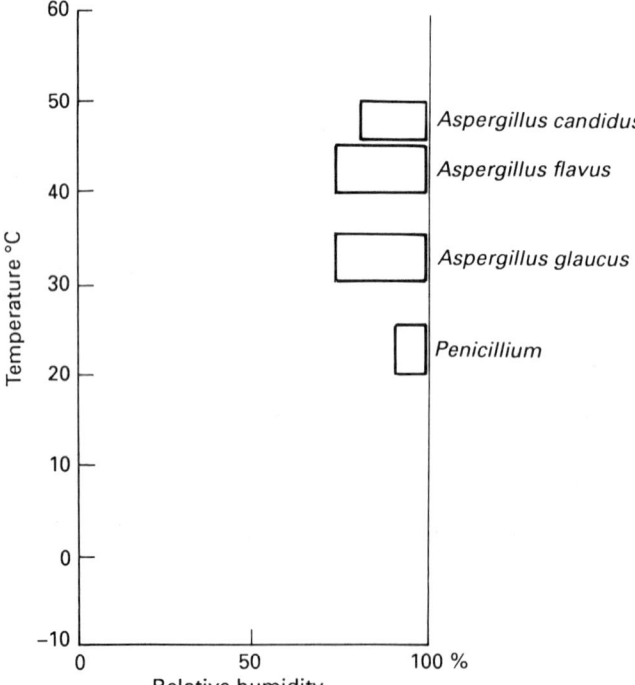

Figure 11.15 Optimum conditions for some grain moulds

the storage container, and occasionally generate toxins. The spores of the hundred thousand different kinds of fungi are continuously present around us and more than 150 have been identified from stored grain. The concentration of such spores is usually particularly high in the vicinity of food warehouses and mills. The small size, simple structure and resilience of the spores means that the only preventive action is to avoid those environmental conditions under which the spores would germinate[20, 21].

The stored food needs to be in an atmosphere around a minimum relative humidity of 70% before germination occurs. Some fungi require even higher humidities. The rate of contamination depends upon the abundance of the initial spores present and on the temperature. Typical values are given in *Figure 11.15*.

Once a mould starts to grow then moisture is one of its metabolic products which can moisten the surrounding food and enhance the rate of development. When this occurs in quantity, then the metabolic heat raises the local temperature and accelerates the process. In spoiling grain, the

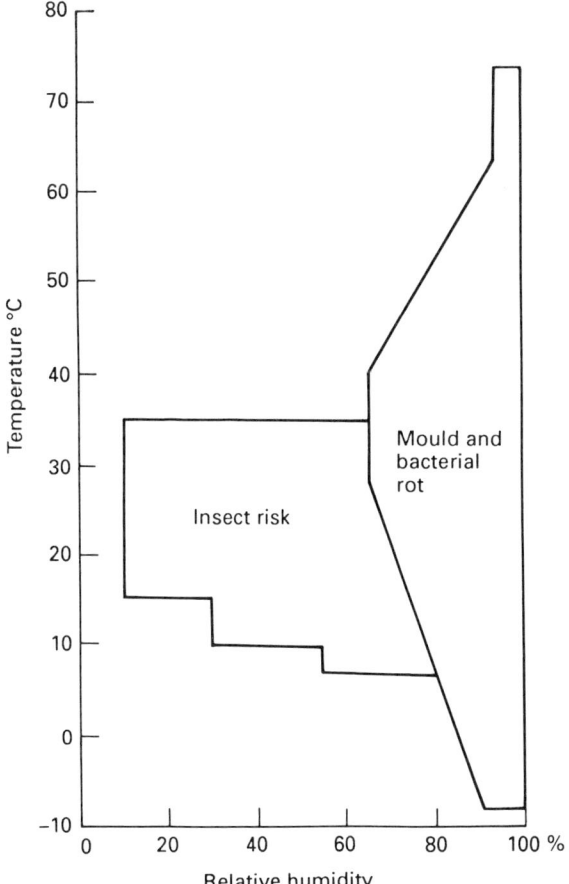

Figure 11.16 Spoilage of stored grain

deterioration by the mould would raise the store temperature to 55°C and enhance the local humidity to the point approaching saturation. In these conditions the thermophilic bacteria multiply and the food is spoiled beyond recovery.

11.8 Practical aspects

Dry food is always at less risk of spoilage than moist food. The equilibrium moisture content of food is uniquely linked to the ambient relative humidity. Low relative humidities are therefore desirable. However, some insects can survive even in very dry food and therefore the critical factor for dry food becomes the temperature. At low temperatures insects do not reproduce or develop quickly and tend to live short lives.

As the relative humidity increases beyond 70%, the risk of spoilage from a wide range of moulds increases, with the added risk that the metabolic heat and moisture release will raise the temperature and moisture content locally. This can then lead to bacterial rot and unrecoverable loss (*Figure 11.16*).

The laboratory experiments described here were all done in small batches under closely controlled conditions. In practice, food stores can be enormous, containing many thousands of kilograms. They will not be of uniform temperature. Since the water vapour pressure will be approximately constant throughout the store, variations in temperature will automatically create variations in relative humidity. Higher temperatures will mean lower relative humidities and vice versa.

Other incidental advantages of a drier food include freer flowing properties when handling and a reduction of 25% in fumigant treatment if the relative humidity is below 65% r.h.[22].

TABLE 11.1 Common names and their corresponding scientific names for food pests

Common name	Scientific name	Family
Confused flour beetle	*Tribolium confusum* Duval	Tenebrionidae
Flat grain beetle	*Cryptolastes pusillus* (Schonherr)	Cucujidae
Flour mill beetle	*Crytolestes turcicus* (Grouv.)	Cucujidae
Grain mite	*Acarus siro* L.	Acaridae
Grain or granary weevil	*Sitophilus granarius*	Curculionidae
Lesser grain borer	*Rhyzopertha dominica* (F)	Bostrichidae
Red flour beetle	*Tribolium castaneum* (Herbst)	Tenebrionidae
Rice weevil	*Sitophilus oryzae*	Curculionidae
Rusty grain beetle	*Cryptolestes ferryineus* (Stephen)	Cucujidae
Saw toothed grain beetle	*Oryzaephilus surinamensis* (L.)	Cucujidae

(Cotton and Wilbur, 1975)[5].

References

1 Harrington, J.F. Seed Storage and Longevity. Vol. 3. *Seed Biology* (ed. T.T. Kozlowski). Academic Press, New York, 1972
2 Uvarov, B.P. 'Insects and climate'. *Trans Entomol Soc London,* **79**, 1–232, 1931
3 Christensen, C.M. (ed.). *Storage of Cereal Grains and their Products.* American Association of Cereal Chemists, Minnesota, USA, 1974
4 Cotton, R.T. *Pests of Stored Grain and Grain Products.* Burgess Publishing Co., Minneapolis, USA, 1963
5 Cotton, R.T. and Wilbur, D.A. 'Insects', Chapter 4, pp. 194–231) in *Storage of Cereal Grains and their Products* (ed. C.M. Christensen). American Association of Cereal Chemists, Minnesota, 1974
6 Milner, M. and Geddes, W.F. 'Respiration and heating', In: *Storage of cereal grain and their products,* Chapter 4, pp. 152–207 (eds J.A. Anderson and A.W. Alcock). American Association of Cereal Chemists, Minnesota, USA, 1954
7 Ford-Robertson, J. *Revise Biology.* Charles Letts & Co., London, 1981
8 Buxton, P.A. 'The law governing the loss of water from an insect'. *Proc R Entomol Soc,* **6**, 27–31, 1931
9 Ramsey, J.A. 'The evaporation of water from the cockroach'. *J Exp Biol,* **12**, 373–383, 1935
10 Gunn, D.L. 'The temperature and humidity relations of the cockroach (*Blatta orientalis*), *J Exp Biol,* **10**, 274–285, 1933
11 Cotton, R.T. 'Insects', Chapter 5 in *Storage of Cereal Grains and their Products* (eds J.A. Anderson and A.W. Alcock). American Association of Cereal Chemists, Minnesota, USA, 1954
12 Howe, R.W. 'The effects of temperature and humidity on the oviposition rate of *Tribolium casteneum*'. *Bull Entomol Res,* **53**(2), 301–310, 1962
13 Mourier, H. and Winding, O. *Collins Guide to Wild Life in House and Home.* Collins, London, 1975
14 Radionov, Z.E. 'Conditions for a mass development of grain mites'. *Zool Xh,* **16**(3), 511–546, 1937. Abstract in *Rev Appl Entomol (A),* **26**, 96
15 Kozulina, O.V. 'The effect of humidity and temperature on the development of the eggs of grain mites'. *Uchem Zap mosk gosud Univ,* **42**, 179–194, 1940. Abstract in *Rev Appl Entomol A,* **31**, 71–72, 1950
16 Solomon, M.E. 'Behaviour of Tyrogliphid mite populations in stored grain and flour'. *Ann Appl Biol,* **31**(1), 81, 1947
17 Cunnington, A.M. 'Physical limits for complete development of the grain mite *Acarus siro* in relation to its world distribution'. *J Appl Ecol,* **2**, 295–306, 1965
18 Blythe, M.E. 'Some aspects of the ecological study of the house dust mites'. *Br J Dis Chest,* **70**, 3–31, 1976
19 Maunsell, K., Wraith, D.G. and Cunnington, A.M. 'Mites and house dust allergy in bronchial asthma'. *Lancet,* 1267–1270, June 1968
19a Wraith, D. G., Cunnington, A. M. and Seymour, W. M. 'The role and allergenic importance of storage mites in house dust and other environments'. *Clinical Allergy,* **9**, 545–561. 1979
20 Semenicek, G., 'Microflora', Chapter 3 in *Storage of Cereal Grains and their Products* (eds J. Anderson and A.W. Alcock). American Association of Cereal Chemists, Minnesota, USA, 1954
21 Christensen, C.M. and Kaufman, H.H. 'Microflora', Chapter 4 in *Storage of Cereal Grains and their Products* (ed. C.M. Christensen). American Association of Cereal Chemists, Minnesota, USA, 1974
22 Anderson, J. and Alcock, A.W. (eds). *Chemical Control of Stored Grain Insects.* American Association of Cereal Chemists, Minnesota, USA, 1954

Further reading

Baker, E.W. and Wharton, G.W. *An Introduction to Acarology (Mites).* New York, 1952
Birch, L.C. 'Experimental background to the study of the distribution and abundance of insects. I. The influence of temperature, moisture and food on the innate capacity for increase of three grain beetles'. *Ecology,* **34**, 698–711, 1953

Cotton, R.T. and Wilbur, D.A. 'Insects', Chapter 4, pp. 194–231 in *Storage of Cereal Grains and their Products* (ed. C.M. Christensen). American Association of Cereal Chemists, Minnesota, 1974

Edney, E.B. *The Water Relations of Terrestrial Arthropods.* Cambridge University Press, 1957

Good, N.E. 'Biology of the flour beetles *Tribolium confusum* and *T. ferrugineum*'. *J Agric Res,* **46**, 327–334, 1933

Hinton, H.E. and Corbet, A.S. *Common Insect Pests of Stored Products, a Guide to their Identification.* British Museum (Natural History) Economics Series 5, London, 1955

Howe, R.W. 'Notes on the biology of *Trogoderma Versicolor* (Col., Dermestidae)'. *Entomologists Monthly Magazine,* **88**, 182–184, 1952

Hughes, A.M. *The Mites of Stored Food and Houses.* Ministry of Agriculture, Fisheries and Food Technical Bulletin No. 9, 2nd edition. HMSO, 1976

Johnson, G.G. 'Insect survival in relation to the rate of water loss'. *Biol Rev,* **17**, 151–177, 1942

Mellanby, K. 'The evaporation of water from insects'. *Biol Rev,* **10**, 317–333, 1935

Park, T. and Davis, M.B. 'The fedundity and development of the flour beetles *Tribolium confusum* and *T. castaneum* at three constant temperatures'. *Ecology,* **29**, 368–374, 1948

Roth, L.M. and Willis, E.R. 'Hygroreceptors in adults of *Tribolium*'. *J Exp Zool,* **116**, 527–570, 1951

Sinha, R.N. 'Seasonal changes in mite populations in rural granaries in Japan'. *Ann Entomol Soc Am,* **61**(4), 938–949, 1968

Chapter 12

Food storage

12.1 Introduction

Food storage techniques have changed rapidly over recent years. The driving force has been the centralization of production, distribution and retailing. For such a system to be effective very high quality control is needed at all stages. The weakest link is still the domestic kitchen, even though most of the households in the Western world have a domestic refrigerator. It was only in 1989 that the first domestic refrigerator with precision temperature control in different compartments became available, and the unit now includes one chamber with controlled humidity.

It is only in recent years that the keeping qualities of many foods have been understood. Originally the risk of spoilage was loosely linked to moisture content. Higher moisture content foods were at higher risk. The concept of available water rather than total water content removed many anomalies[1,2]. This has introduced the concept of water activity a_w.

We will introduce this subject and then turn to the major food categories of bakery products, confectionery, meat and finally fruit and vegetables.

12.2 Water activity and equilibrium relative humidity

Many substances dissolve readily in water. These substances are called solutes and the liquid in which they dissolve is termed the solvent. The liquid is then a solution. This solution no longer behaves as pure water. The freezing point is lowered, the boiling point is raised and most importantly the water vapour pressure is lowered.

For an ideal solution, in which the forces acting between the different types of molecule are the same, Raoult's Law states that the relative lowering of the vapour pressure of a solvent is related to the proportion of molecules dissolved in the solvent. However, the molecules of different compounds are of different weight and therefore to convert molecules to weight we need to work in units of gram molecules. This is the molecular weight in grams and is termed a 'mole'. Equal moles contain equal numbers of molecules.

Raoult's Law may be expressed as:

$$\frac{\text{vapour pressure of the solution } p}{\text{vapour pressure of pure water } p_0} = \frac{\text{moles of solvent}}{\text{moles of solute (e.g. sugar)} + \text{moles of solvent}}$$

If we dissolve one mole of solute into one litre of water, which is 55.5 moles, then the expression becomes:

$$\frac{p}{p_0} = \frac{55.5}{1 + 55.5} = 0.982$$

i.e. the vapour pressure of the solution is 98.2% of that of the saturated vapour pressure of water at the same temperature. The equilibrium relative humidity in the air above this solution will be 98.2%.

In practice this means that knowledge of the moisture content of food does not give any guide to the availability of water and hence to its keeping properties. For microbiological purposes the critical factor is the free moisture. The amount of free moisture is determined by the amount of solute present. Both the solute and the microorganisms compete for the water. The availability of water is defined by the water activity a_w. When all the water is available, $a_w = 1.0$. As the amount of solute increases in the water, so the value falls.

Water activity is numerically equal to the equilibrium relative humidity when expressed as a fraction, i.e. $a_w = 0.982$ for the 1 mole solution described above[1,2].

In a given weight of substance the number of moles present will be inversely proportional to the molecular weight of the compound. This number will be even bigger if the molecule dissociates into ions when dissolved. Sodium chloride dissociates so well that it effectively halves its molecular weight and doubles its effect.

The effect of three common solutes, salt, glycerol and sucrose, in cooking processes is illustrated in *Figure 12.1*[3].

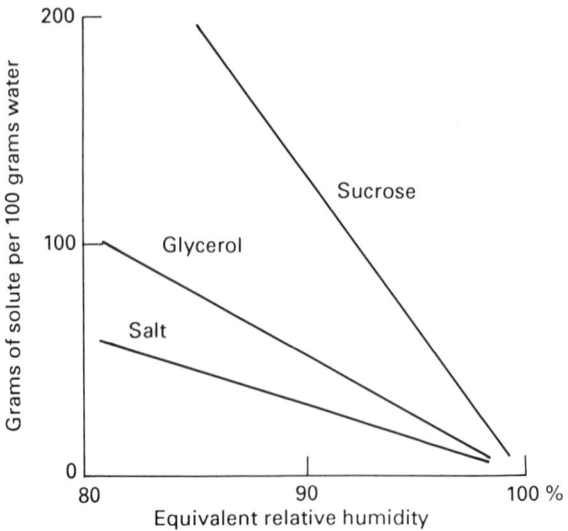

Figure 12.1 Influence of solutes on water vapour pressure

12.3 Bakery products

Foodstuffs can experience a wide range of changes during storage. These include physical changes such as drying or moistening, chemical changes such as oxidation, continued enzyme activity and microbiological contamination[4]. The most rapid spoilage is caused by microbiological factors. All microbiological activity requires water for growth and one advantage of using sugar, for example, as a food preservative, is that it lowers the water activity a_w and the equilibrium relative humidity. The critical factor is a_w, not total moisture content, and the microorganisms have to compete with the solute molecules for the water they need.

A range of equilibrium relative humidities for domestic foods is illustrated in *Figure 12.2*[3]. Dry biscuits and cream crackers represent the most dry and breadcrumbs the most moist. Dry biscuits rarely experience microbiological attacks because they would first need to increase their actual moisture content to 10–13% to reach an equivalent humidity of 70% and in this condition would have an unacceptable eating quality.

Appropriate ranges of activity for microorganisms are also shown in *Figure 12.2*. Most bacteria only grow above 94–95% r.h., although *Staphylococcus aureus*, often associated with food poisoning, will grow extremely slowly down to 86% r.h. Normal yeasts can grow down to 88% r.h., although there are certain strains termed osmophilic yeasts which can

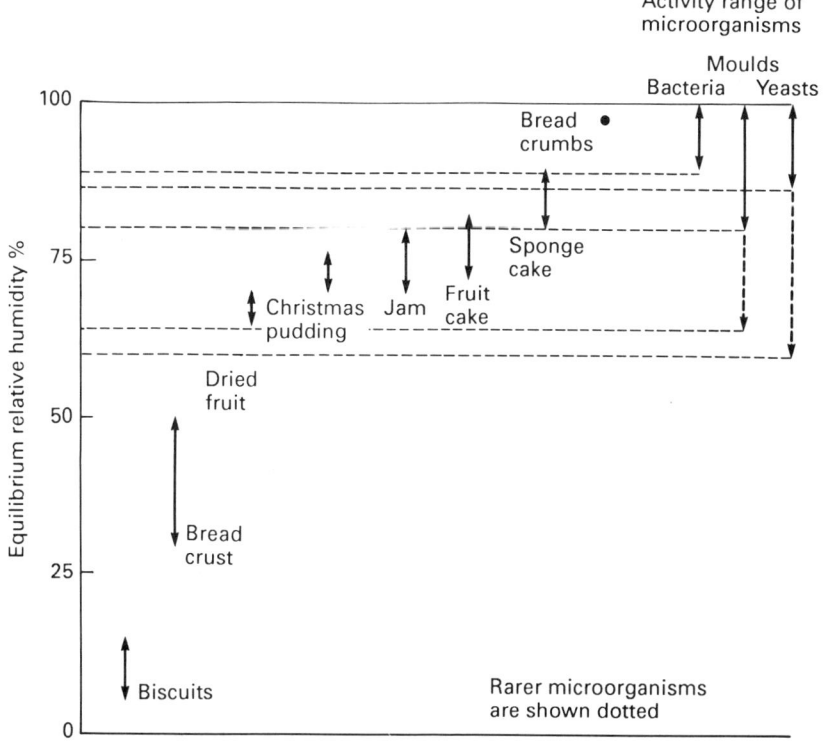

Figure 12.2 Approximate equilibrium relative humidity for bakery products

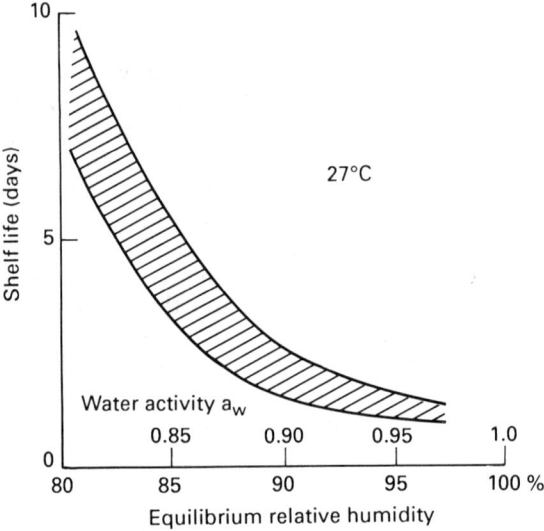

Figure 12.3 Mould-free shelf life for cake varies with its relative humidity

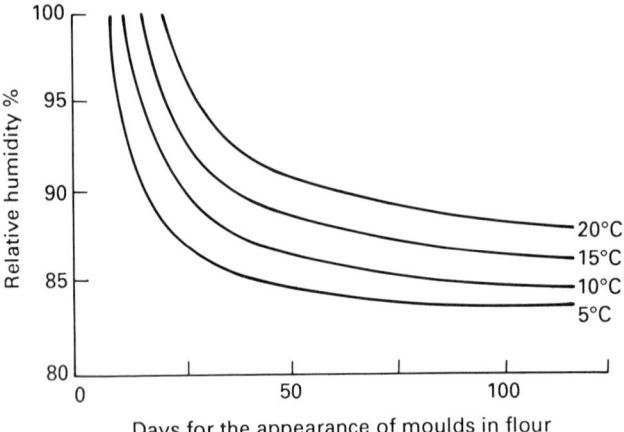

Figure 12.4 The time taken before mould appears in flour is a function of ambient relative humidity and temperature

grow in high concentrations of sugar and syrup. Such yeasts can attack jam, marzipan and marshmallow. Moulds are the most water-tolerant of the microorganisms and many can grow down to 80% r.h., with a few such as *Aspergillus glaucus* developing very slowly down to 65% r.h.

The mould-free shelf life of cake is determined mainly by its equivalent relative humidity. The relationship between the two is illustrated in *Figure 12.3* for wrapped cakes at a temperature of 27°C, which would favour the rapid development of mould.

For simple raw materials such as flour, the onset of mould is related to temperature and relative humidity. The relationship is illustrated in *Figure 12.4*[5].

12.4 Confectionery

Confectionery contains a wide range of different ingredients which include sugar in any of its many varieties, such as sucrose, dextrose and invert syrups, chocolate, nuts, fruit, milk products, butter, eggs, flour, starch and colouring. There is no specific critical storage temperature but in general the lower the temperature the longer the storage life, providing moisture can be controlled. The lower temperature inhibits the degeneration of the fats and proteins which eventually develop 'off' flavours. Confectionery which contains nuts or fruit are possible hosts for insect eggs, and insects are rendered inactive below 9°C. However, a few weeks exposure to −18°C is needed to destroy insect life.

Moisture has two kinds of effects, surface damage and bulk modification. The surface damage is usually attributed to transient condensation caused by bringing the product out of a cold store and exposing it to air with a higher dew point and hence incurring condensation. This can cause sugar blooming on chocolates which imparts an unattractive grey colour. It can be caused by a deposition of crystallized sugar on the surface following condensation.

The changes in the bulk of the product are the result of moisture loss or gain. If sugared products lose their moisture they become hard. The critical humidity varies widely with the product and the hygroscopicity of the ingredients. In general the high water activity products such as marshmallows, gum drops, jelly beans and fudge should be stored at 60–65% r.h. The medium ones such as caramels and nougat should be at 50–55% r.h. and the low activity products such as chocolate bars and barley sugars should be stored at 45% r.h. or lower. Products using a high proportion of invert syrup need an even lower relative humidity because of the hygroscopic nature of the syrup. The influence of temperature on storage life is shown in *Figure 12.5*. Deviations from optimum conditions of

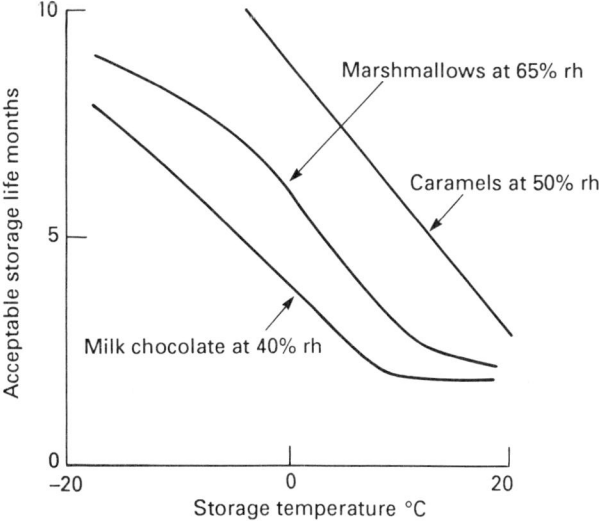

Figure 12.5 The satisfactory storage of confectionery

moisture are more permissible at lower temperatures because the speed of moisture take-up is more determined by the actual water vapour pressure than by the relative humidity. Stickiness caused by the relative humidity being 10% too high may take one day at 20°C or several weeks at 0°C.

12.5 Meat and its products

Fresh lean meat has a water activity of 0.99, i.e. an equilibrium relative humidity of 99%, and hence is very similar to pure water. All meat products have a lower water activity because shelf life is dramatically improved with lower values. This is usually achieved by the removal of water or by the addition of salts or both. The range of water activities for typical meat products is given in *Table 12.1*[7]. The lowest value is for the salami which ranges from 0.65 to 0.96.

TABLE 12.1 Water activity for meat products

Product	Minimum water activity	Maximum water activity	Modal value of water activity
Fresh meat	0.98	0.99	0.99
Bologna sausage	0.93	0.98	0.97
Liver sausage	0.95	0.97	0.96
Black pudding	0.86	0.97	0.96
Raw ham	0.80	0.96	0.92
Dried beef	0.80	0.94	0.90
Salami	0.65	0.96	0.91

(Leistner and Rodel, 1976)[7].

There are two types of microbiological spoilage which affect meat products. The first is the multiplication of bacteria which break down the protein to create visible spoilage and putrefaction and create a distinct smell. These are not normally dangerous to humans, although they clearly serve as indicators of poor food control. The second is the multiplication of food-poisoning bacteria. These are insidious because food grossly contaminated by such organisms does not have a different appearance, smell or taste from that of good food[8].

The three international food-poisoning bacteria are *Salmonella*, *Staphyloccus* and *Clostridium welchii*. *Salmonella*, usually from animal excreta, introduces an infection when eaten in large quantities. *Staphylococcus* is readily found on skin and in the nose, and when it multiplies it produces a toxic metabolite. *Clostridium welchii* will only grow in the absence of oxygen and its widespread presence is attributed to its hardy spores.

Studies on *Staphylococcus aureus* showed that the bacteria multiplication rate declined progressively with values of water activtity below 0.995. Total suppression of all strains occurred at a water activity of 0.84[9]. This is illustrated in *Figure 12.6*.

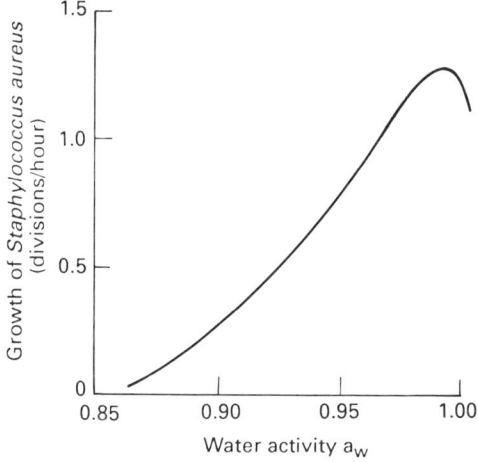

Figure 12.6 Growth of the food-poisoning bacteria *Staphylococcus aureus*

In practice, the multiplication of microorganisms on or in meat products depends upon a wide range of factors. Temperature and water activity are the two most important but these are strongly influenced by the pH, the nitrite content and the competitive flora. German studies suggest a simple grouping which provides a good guide to minimizing the risk of contamination from both food spoilage and food poisoning. This guide is illustrated in *Table 12.2*.

TABLE 12.2 Storage categories for meat products

Category	Criteria	Temperature
Storable	$a_w < 0.95$ and pH < 5.2 or $a_w < 0.91$ or pH < 5.0	No refrigeration required
Perishable	$a_w < 0.95$ or pH < 5.2	$< 10°C$
Very perishable	$a_w > 0.95$ and pH > 5.2	$< 5°C$

(Leistner and Rodet, 1976)[7].

12.6 Fresh vegetables, fruits and flowers

Cut plants are still living tissues, albeit living for survival rather than growth. Their useful shelf life is determined by any one of three factors. The first is the length of time the metabolic processes can keep the plant alive. This is a function of the amount of nutrient available within the plant and the rate at which this nutrient is consumed. This process is termed respiration. The second is moisture content of the plant. Most plants are 80–90% water by weight. Loss of water means an unattractive appearance because the plant has wilted or the fruit shrivelled. This moisture loss is termed transpiration. The final factor is spoilage, which is usually fungal attack. Let us examine the first two factors in more detail.

(a) *Respiration*

This is the most important metabolic process by which the oxygen in the air continues to be combined with the carbon of the plant tissue, chiefly in the form of sugars, to form various decomposition products leading ultimately to carbon dioxide and water vapour.

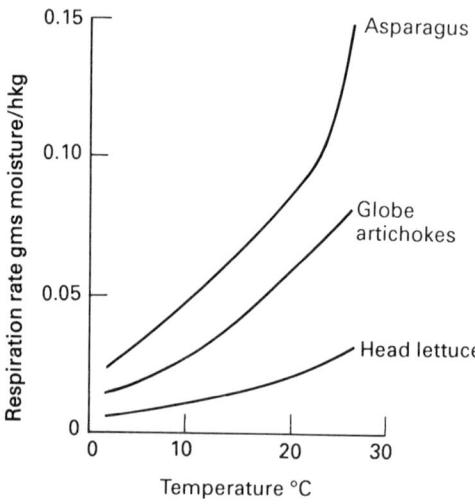

Figure 12.7 Respiration rates for vegetables

The process can be grossly simplified to the oxidation of hexose sugars:

sugar + oxygen = carbon dioxide + water vapour + energy
$$C_6H_{12}O_6 + 6O_2 = 6CO_2 + 6H_2O + 673\,k\,cal$$
$$192\,g + 192\,g = 284\,g + 108\,g + 673\,k\,cal$$

The energy is released in the form of heat. The actual amount varies with the plant and increases with increasing temperature up to 40°C. The rate of reaction is governed by temperature, doubling or tripling with every 10°C rise in temperature. There is a very wide range in respiration rates for different plants at the same temperature. Illustrations of this reflected in the moisture release rate are given in *Figure 12.7* for vegetables and in *Figure 12.8* for fruit. The water released by respiration is normally one-tenth of that lost by evaporation. The carbon dioxide concentration will tend to build up in the storage room and monitoring is needed to avoid high concentrations of 4%. This is particularly important if plants such as lettuce, which degrade in the presence of carbon dioxide, are stored in the same room.

The controlling factor is the depletion of the sugar reserves of the plant. Refrigeration slows down the rate of use of this sugar and prolongs life dramatically.

(b) *Transpiration*

Evaporation from living tissue is termed transpiration. The driving force for evaporation is the vapour pressure difference between that associated

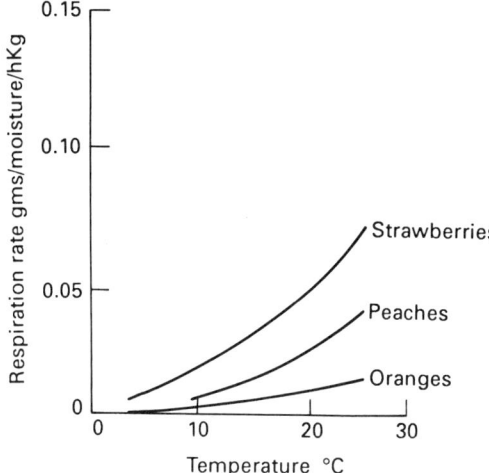

Figure 12.8 Respiration rates for fruit

with the water in the plant and that of the ambient environment. The resistance to evaporation is the permeability and area of the plant skin. This varies widely for different plants.

The critical factor in moisture loss is the physical appearance of the plant. Shrivelled plants are unacceptable to the prospective purchaser. The importance of this moisture loss varies widely with the commodity, and illustrations are given in *Figure 12.9*. Grapes are amongst the most sensitive because they lose their full rounded appearance when only 1% of water is lost[11]. Apples look shrivelled when 5–7% of the weight is lost[12, 13]. Laboratory measurements on stored Golden Delicious apples show how

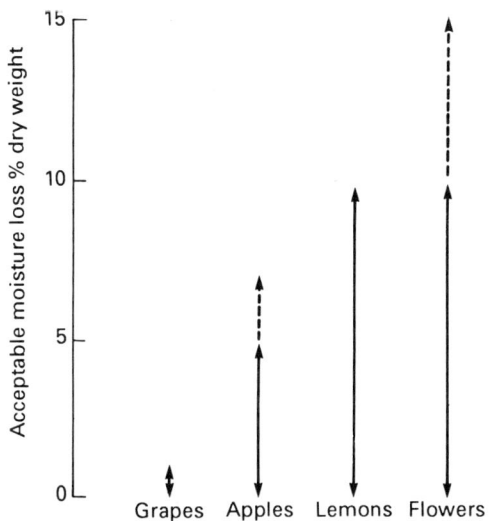

Figure 12.9 Moisture loss before unacceptable appearance of shrivelling

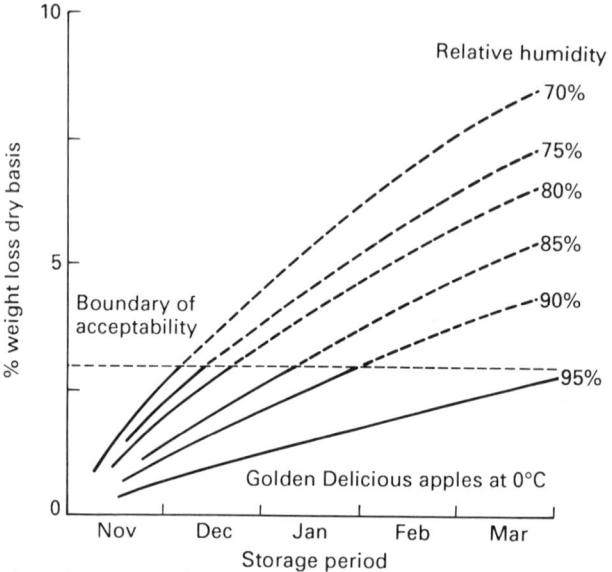

Figure 12.10 Weight loss of apples at 0°C as a function of relative humidity (2% loss assumed during picking and transport)

useful life can be enhanced by storing at a high ambient relative humidity of 95%. This is illustrated in *Figure 12.10*. An arbitrary 2% additional weight loss is attributed to losses during picking and transporting to the laboratory. Shrivelling also occurs selectively. In grapes the stem shrivels first but this does not matter. In some flowers the petals shrivel quickly and therefore it is essential that the relative humidity is always above 87%[14].

There are three methods of inhibiting this moisture loss. The first is to store at a high relative humidity so that the driving force for evaporation is low. This has two practical implications. The first is that the risk of spoilage by mould increases. The second is that the precision and uniformity of temperature control within the store becomes progressively more important as the storage temperature is lowered and the relative humidity increases. Non-uniformity of temperature within the store can readily lead to local condensation problems which may lead to deterioration of both the produce and the building.

The second method is to package the produce in perforated polyethylene wrappers or container liners. This slows down the rate of moisture loss and tends to retain the respiration moisture generated by the plant.

The third method is to lightly wax the outer skin. This is often used for cucumbers.

Precise guidelines on ideal storage conditions are difficult to produce because of the wide variety of species within each category. The guidelines also have to allow for present-day technology. As the technology of thermal control and microbiological control becomes more refined, the guidelines are able to approach the ideal conditions, particularly saturation relative humidity for those plants which need it[10, 15].

In general, the controlling physical factor for shelf life is temperature. Assessments of plant quality show a fairly uniform decline with time but the rate of change is determined by the ambient temperature. Celery, for example, degrades from 'perfect' to 'acceptable' in one day at 20°C but takes 27 days to reach the same state at 0°C. This is illustrated in *Figure 12.11*. However, each plant has its own temperature/time relationship. This is shown in *Figure 12.12*. There are anomalies. Cucumber, for example, has an optimum storage temperature of 14°C, where it can have a shelf life of 10 days. If it is stored at 3°C this is reduced to three days. This is also illustrated in *Figure 12.12*. Lemons develop pitting and staining

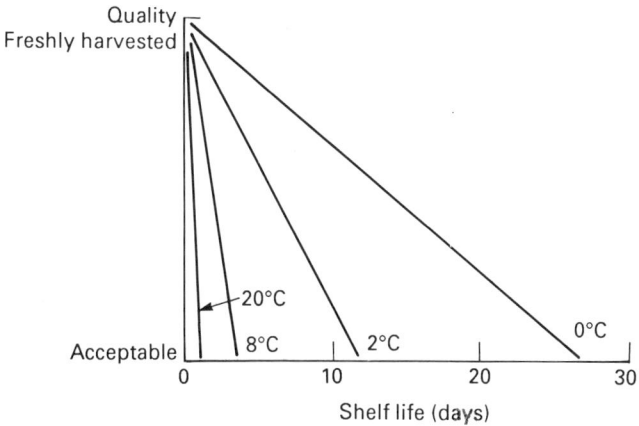

Figure 12.11 The effect of temperature on the shelf life of celery

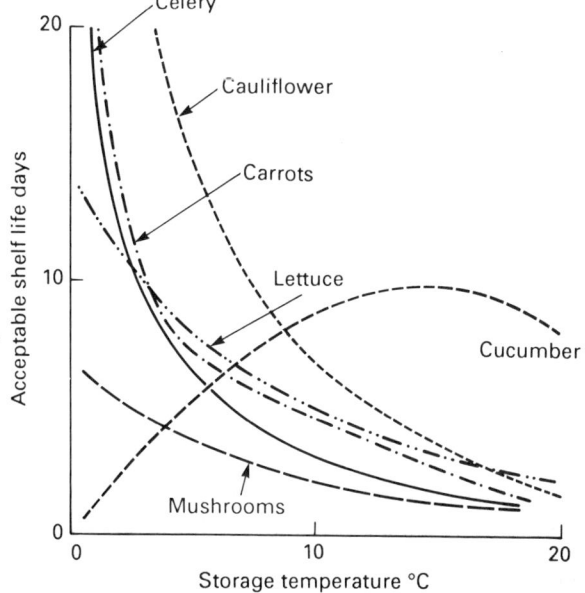

Figure 12.12 The influence of tempereature on vegetable shelf life

between segments if stored below 14°C. Similarly, some fruits have special sensitivities to relative humidity. The peel of grapefruit tends to break down below 85% r.h. and pears lose moisture very rapidly at low relative humidities.

Guidelines for the ideal storage conditions for fruit and vegetables are given in *Table 12.3*.

TABLE 12.3 Ideal storage conditions for fruit and vegetables (Bioteknisk Inst. Denmark 1987)[15]

Produce	Ideal temperature (°C)	Ideal relative humidity (%)	Storage life
Fruit			
Apples	0–6	95–100	1–2 weeks
Avocados (packaged)	7–13	85–90	6–7 days
Bananas (ripe)	12–14	95–100	3–6 days
Clementines	0–4	85–90	7–10 days
Grapefruit	10–15	85–90	2–4 weeks
Grapes	0	95–100	135 days
Lemons (ripe)	11–15	85–90	1–2 weeks
Oranges	2–5	90–95	2–3 weeks
Pears (packaged)	0	95–100	5–7 days
Pineapple (ripe)	8–10	90–95	5–7 days
Strawberries	0	95–100	5 days
Vegetables			
Brussels sprouts	−1–0	95–100	20 days
Cabbage	0	95–100	200 days
Carrots	0	95–100	27 days
Cauliflower	0	95–100	42 days
Celery	0	95–100	27 days
Cucumber	7–14	95–100	10 days
Leeks	−1–0	95–100	41 days
Lettuce	0	95–100	14 days
Mushrooms	0	95–100	7 days
Onions	−2–0	75–85	250 days
Potatoes	4–6	95–98	240 days
Tomatoes	11–14	75–80	13 days

12.7 Conclusions

Water is the main constituent in many foods. It accounts for 80–90% by weight of fruits and vegetables, and between 70–80% of meat and fish. The control of this water is critical. Food storage involves the delicate balance between maintaining sufficient water in such foods to maintain appearance, texture and freshness, while discouraging microbiological contamination.

Empirical techniques such as curing, smoking, salting and jamming have only recently been put on a scientific basis. The chemists have talked to the microbiologists and have recognized that moisture content alone is not a sufficient guide to keeping qualities. The critical factor is the availability of the water. The water vapour pressure associated with solutions is quite different from that associated with pure water. The addition of solutes such as salt or sugar lowers the effective pressure. Any microorganisms, therefore, have to compete with solute for available water. The new term to quantify this effect is called water activity a_w. It is numerically equivalent to the equilibrium relative humidity above the product. Low water activity a_w means a low equilibrium relative humidity. It is normally associated with a strong solution and has a long shelf life because microorganisms find it difficult to extract the bound water which is there.

Much is also now known about the microorganisms which attack food. Bacteria will only survive in high free moisture conditions usually represented by ambient relative humidities above 90%. There are two types. One degrades the food and leaves it looking unsightly and smelly. These are usually harmless to man. The second contamines the food in a way which does not alter the taste, appearance or flavour. The food poisoning bacteria belong to this second group and can create serious infection or unpleasant toxic poisoning, depending on the strain. The problems are usually associated with food being heated to a lukewarm temperature for a long time. In these circumstances the microorganisms will colonize the product and when eaten will overwhelm the natural defences of the individual.

Mould and yeasts can develop in much lower relative humidities down to typically 70%. They are important in spoilage although it is only recently that there has been concern over possible health effects. Epidemiological studies have linked the eating of mouldy nuts with liver problems. Researchers are actively exploring the toxicity of the metabolites of mould. To date, only a small proportion of moulds associated with stored foods appear to produce toxins. Only 10% of *Aspergillus* and *Penicillium* strains found on cereals generated toxins harmful to mice. Some 16% cultured from aged cured meat produced a substance which was toxic to chicken embryos. Fortunately the toxins appear to be produced when the mould is thriving rather than when it is struggling to grow.

The most perishable foods are those which are still living, the fruits and vegetables. The ultimate factor determining shelf life is the amount of nutrient stored within the plant and the rate at which this is being used. This is strongly linked to temperature. Moisture plays an important role in physical appearance. If too much water is lost then the food will look shrivelled and unattractive and will not sell. Moisture also plays a critical role in the multiplication of microorganisms. In practice, there are many other factors, such as care in handling, avoiding bruising, careful planned stacking and good warehousing services which ensure uniform conditions throughout the store.

The reorganization of the food distribution chain within Britain from large production centres, through refrigerated transport and into carefully supervised stores is introducing new higher quality control standards.

References

1 Grover, D.W. 'The keeping properties of confectionery as influenced by its water vapour pressure'. *J Soc Chem Ind*, **66**, 201–205, 1947
2 Scott, W.J. 'Water relations of food spoilage micro-organisms. In *Advances in Food Research*, **7**, 83–123, 1957

3 Cooper, R.M., Knight, R.A., Robb, J. and Seiler, D.A.L. 'The equilibrium relative humidity of baked products with particular reference to the shelf life of cakes'. Flour Milling and Baking Research Association. Report No. 19. September 1968, 38pp

4 Acker, L. 'Enzymic reactions in foods of low moisture content. *Adv Food Research,* **11,** 263–330, 1962

5 Barton-Wright, E.C. and Tomkins, R.G. 'The moisture content and growth of mould in flour, bran and middlings. *Cereal Chemistry,* **17,** 332–342, 1940

6 American Society of Heating, Refrigerating and Air Conditioning Engineers. Applications Handbook 1986

7 Leistner, L. and Rodel, W. Inhibition of microorganisms in food by water activity, pp. 219–238, in *Inhibition and Inactivation of Vegetative Microbes* (eds F.A. Skinner and W.B. Hugo). Academic Press Ltd, London, 1976

8 Hobbs, B.C. 'Food poisoning and food hygiene'. Edward Arnold, 1974

9 Troller, J.A. and Christian, J.H.B. 'Water activity and food'. Academic Press, New York, 1978

10 Lutz, J.M. and Hardenburg, R.E. 'The commercial storage of fruits, vegetables and florist and nursery stocks'. US Department of Agriculture, Agricultural Handbook 66, Revised 1968

11 Allen, F.W. and Pentzer, W.T. 'Studies on the effect of humidity on the cold storage of fruits'. *Amer Soc Hort Sci Prod,* **33,** 215–223, 1936

12 Smith, W.H. 'Evaporation of water from apples in relation to temperature and atmospheric humidity'. *Annal Appl Biol,* **20**(2), 220–235, 1933

13 Christopher, E.P., Pieniazek, S.A. and Shutak, V. 'Transpiration of apples in cold storage. *Amer Soc Hort Sci Proc,* **51,** 114–118, 1948

14 Hitchcock, A.E. and Zimmerman, P.W. 'Effect of chemicals, temperature and humidity on the lasting quality of cut flowers'. *Amer J Bot,* **16,** 433–440, 1929

15 Bioteknisk Institut Kolding Denmark. Handbog om frugt und gronsager. Akademiet for de Tekniske Videnskaber, 1987

Further reading

Ames, A. 'The temperature relations of some fungi causing storage rots'. *Phytopathology,* **5**(1), 11–19, 1915

Ayerst, G. 'Water activity – its measurement and significance in biology'. *Internat Biodeterioration Bulletin,* **1**(2), 13–26, 1965

Berg, L. van den and Lentz, C.P. 'Effect of temperature, relative humidity and atmospheric composition on changes in the quality of carrots during storage'. *Food Technology (USA),* **20,** 954–957, 1966

Berry, J.A. and Magoon, C.A. 'Growth of microorganisms at and below 0°C'. *Phytopathology,* **24**(7), 780–796, 1934

Beuchat, L.R. 'Food and beverage mycology'. AVI Publishing Corp. Connecticut, USA, 1978

Brooks, F.T. and Hansford, C.G. 'Mould growths upon cold store meat'. *Brit Mycol Soc Trans,* **8,** 113–142, 1923

Brown, W. 'On the germination and growth of fungi at various temperatures and in various concentrations of oxygen and carbon dioxide'. *Ann Botany (London),* **36,** 257–283, 1922

Christjakoff, F.M. and Bocharova, Z.Z. 'The influence of low temperature on mould'. *Mikrobiologia,* **7,** 498–513, 1938

Chordash, R.A. and Potter, N.N. 'Effects of dehydration through the intermediate moisture range on water activity microbial growth and texture of selected foods'. *J Milk Food Technology,* **35**(7), 395–398, 1972

Cutting, C.L. and Hardy, J.K. 'The influence of temperature of storage on the loss of water from stored material'. DSIR Report of the Food Invest. Board for 1938, pp. 245–251. HMSO, London, 1939

Fields, M.L. 'Fundamentals of food microbiology'. AVI Publishing, Connecticut, USA, 1979

Goos, R.D. and Tschirsch, M. 'Effect of environmental factors on spore germination, spore survival and growth of *Glocosporum Musarum'*. *Mycologia,* **54,** 353–367, 1962

Gorini, F.L. and Zanetti, A. 'La perdita di peso in akune pomacce'. Research on the weight loss in apples and pears. Annali Instituto Sperimentale per la Valorizzazione Tecnologica dei Prodolt. *Agricoli,* **3,** 209–213, 1972

Grower Guide No. 22. 'Cut flowers from bulbs'. Grower Books, London, 1981

Haines, R.B. 'Microbiology in the preservation of animal tissues'. DSIR Food Investigation Special Report No. 45. 88pp, 1937

Lentz, C.P. 'Moisture loss of carrots under refrigerated storage conditions'. *Food Technology*, **20**, 201–204, 1966

Lentz, C.P. and Rooke, E.A. 'Rates of moisture loss of apples under refrigerated storage conditions'. *Food Technol*, **20**, 201–204, 1966

Marin, G. 'The temperature and humidity in cold stores and the loss of water from stored produce'. *Proc Int Inst Refrig*, **48**, 102–107, 1951/2

Platenius, H., Jamison, F.S. and Thompson, H.C. 'Studies on cold storage of vegetables'. Cornell Univ. Agric. Experimental Station, New York. Bulletin 602, 1934

Richardson, K.C. 'Handling food in the home'. *CSIRO (Australia), Food Research Quarterly*, **35**, 25–33, 1975

Riddet, W. 'Butter boxes and mould growth'. *New Zealand Journal of Agriculture*, **53**(3), 129–139, 1936

Romijn, J.G. 'Improved conditioned cheese storage'. 15th International Congress of Refrigeration, Vol. 3, 1189–1198, 1979

Roussel, L. and Vidal, P. 'Relative humidity in cold storage – buffer effect of packaging'. Proc. 10th International Congress on Refrigeration (Copenhagen). Section 5, Paper 18, 6pp. 1959

Sastry, S.K. 'Moisture losses from perishable commodities: recent research and developments'. *Rev Int Froid*, **8**, 343–346, 1985

Schelhorn, M. von. 'Control of microorganisms causing spoilage in fruit and vegetable products'. *Adv Food Res*, **3**, 429–482, 1951

Skinner, F.A. and Hugo, W.B. 'Inhibition and inactivation of vegetative microbes'. Academic Press Ltd, London, 1976

Smart, H.F. 'Microorganisms surviving the storage period of frozen pack fruits and vegetables'. *Phytopathology*, **24**(12), 1319–1331, 1934

Smith, W.L., Miller, W.H. and Bassett, R.D. 'Effects of temperature and relative humidity on the germination of *Rhizopus stolonifer* and *Monilinia fructicola* spores'. *Phytopathology*, **55**(6), 604–606, 1965

Snow, D., Crichton, M.H.G. and Wright, N.C. 'Mould deterioration of feedingstuffs in relation to humidity of storage'. *Ann Appl Biol*, **31**(2), 111–116, 1944

Torrey, G.S. and Marth, E.H. 'Isolation and toxicity of moulds from foods stored in homes'. *J of Food Protection*, **40**(3), 187–190, 1977

Walbeck, W. van, Clademenos, T. and Thatcher, F.S. 'Influence of refrigeration on aflatoxin production by strains of *Aspergillus flavus*'. *Canadian J. Microbiology*, **15**, 629–632, 1969

Weimer, J.L. and Harter, L.L. 'Temperature relations of eleven species of *Rhizopus*'. *Journal of Agric Res (USA)*, **24**(1), 1–40, 1923

Williams, D.B. and Purnell, H.G. 'Spore germination, growth and spore formation by *Clostridium Botulinum* in relation to the water content of the substrate'. *Food Research*, **18**, 35–39, 1953

Zachariah, A.T., Hanson, H.N. and Snyder, W.C. 'The influence of environmental factors on cultural characters of *Fusarium* species'. *Mycologia*, **48**, 459–467, 1956

Chapter 13

Controlling humidity in museums

13.1 Introduction

Museums contain our heritage of valuable works of art and literature, together with our early tools, instruments and machinery. Most of these objects are sensitive to moisture but often in very different ways. For example, the drier conditions most suitable for metal objects would be harmful to books stored there, by making the paper and the binding too brittle.

The advances in air conditioning enable the indoor climate to be filtered and cleaned, and its temperature and moisture controlled. Curators are recognizing how to balance the economics of plant operation with the care of their museum displays.

There are three types of damaging processes influenced by moisture. These are:

(1) *Physical damage.* This is the shrinking and cracking due to changes in the physical size of the fabric of the exhibit as the humidity changes.

(2) *Chemical deterioration.* This is most typified by the corrosion processes which occur in damp conditions.

(3) *Biodeterioration.* This is the destruction of the exhibit by moulds or insects.

Let us consider them in turn.

13.2 Physical damage

Common museum materials such as wood, bone, ivory, parchment, leather, lacquer, textiles, paper and adhesives readily adsorb moisture. Moisture-absorbing materials increase in size as they adsorb moisture. The critical factor influencing the equilibrium moisture content of any material is the relative humidity of the ambient air. The relationship between moisture content and ambient relative humidity at a given temperature is termed the adsorption isotherm. Examples are illustrated in *Figure 13.1*[1].

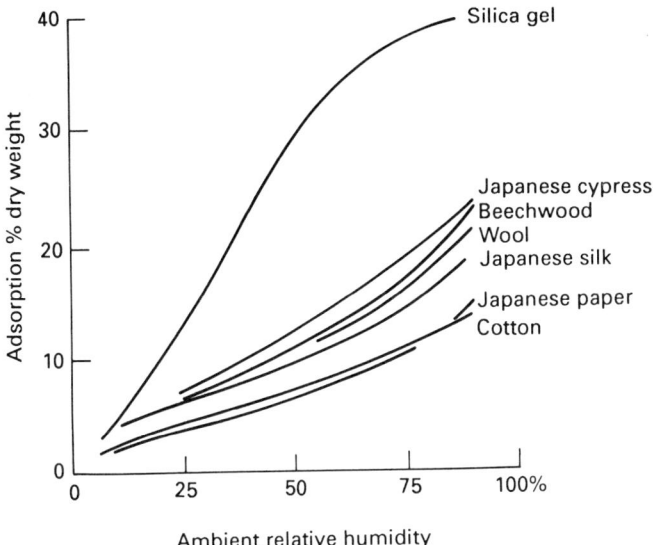

Figure 13.1 Adsorption isotherms for representative materials (Toishi, 1959)[1]

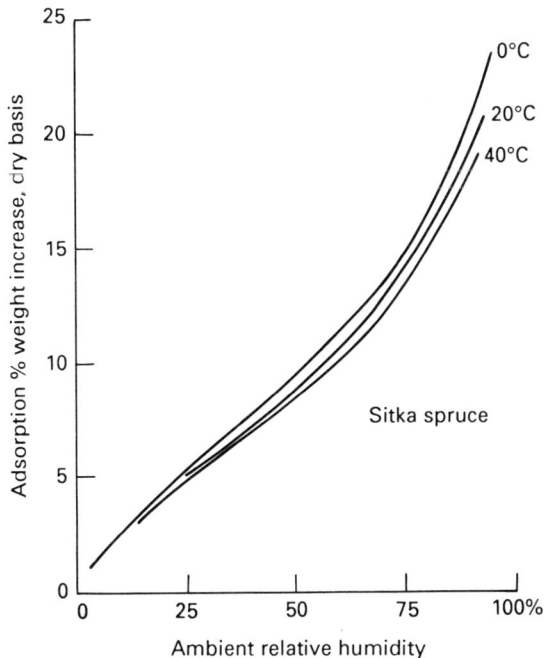

Figure 13.2 The adsorption isotherm changes slightly for different temperatures (Kollman and Cote, 1968)[2]

There is a small effect of temperature which is illustrated for wood in *Figure 13.2*[2]. In practice these figures are a general approximation of the truth because the adsorption isotherm is affected by the pre-treatment and conditioning of the particular material, and the actual equilibrium moisture content is affected by the direction of moisture change. This hysteresis associated with direction of change is illustrated in *Figure 13.3* for a typical wood. The initial drying desorption never recovers fully and the treated timber then cycles within the shaded hysteresis area for varying ambient conditions of relative humidity.

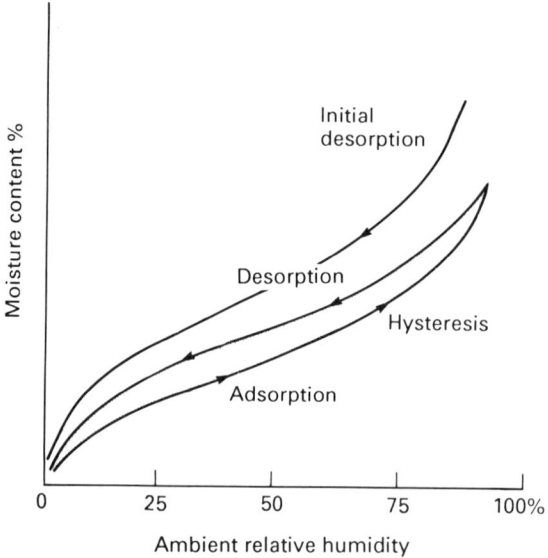

Figure 13.3 Diagrammatic illustration of hysteresis in the adsorption isotherm

Wood deserves special care because it undergoes a large change in size with changing moisture content. The problems can be of three different types. The first and simplest is shrinkage in dry conditions. This leads to furniture becoming loose at the joints or, for example, the axe handle falling out of the axe head. The second is distortion, which can twist doors, picture frames and panelling during transient changes in humidity. The third is splitting of thick timber due to an excessive rate of desorption, usually over a prolonged period of several days or more.

The critical factor in all these effects is the grain direction of the timber. Shrinkage and swelling do not occur uniformly in wood. The greatest dimensional changes due to moisture occur in the direction tangential to the circumferential annual wood rings in the trunk of the tree. Shrinkage radially from the centre of the tree is typically two-thirds of the tangential shrinkage. Axial shrinkage is practically negligible at less than one-twentieth of the tangential shrinkage (*Figure 13.4*)[2].

Typical values are listed in *Table 13.1*. In general the denser woods display greater shrinkage on desorption[2,3]. Tangential swelling which

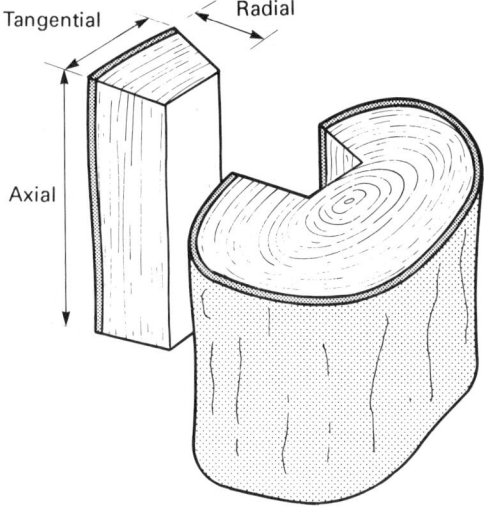

Sketch of a tree trunk

Figure 13.4 Moisture changes in wood affect the tangential dimension most (Kollman and Cote, 1968)[2]

TABLE 13.1 Dimensional expansion in wood on going from an atmosphere at 50% r.h. to 60% r.h.

Species	Tangential expansion (%)	Radial expansion (%)
Mahogany	0.45	0.30
Chestnut	0.50	0.35
Walnut	0.50	0.40
Scots pine	0.50	0.40
Oak	0.70	0.40
Birch	0.85	0.65
Beech	0.90	0.60

Longitudinal shrinkage is approximately one twenty-third of the tangential shrinkage, i.e. 0.02–0.04% for a 10% change in r.h.

Thomson, 1978[3].

occurs when the ambient relative humidity varies from 50% to 60% r.h. varies for the different woods from ½% to almost 1%. Tests on timber which varied in age from less than one year to almost 4000 years did not reveal any influence of age[4].

Transient changes in ambient relative humidity regularly cause distortion and bending of timber. This effect is made worse if the timber has been cut tangentially to the annual rings on the trunk and if only one side of the timber has been painted or varnished. The changes in moisture content occur more quickly from the unprotected side of the timber and make the timber bend[5].

Finally, the speed with which timber responds to changing ambient moisture conditions depends upon the thickness of the piece of timber.

Experiments on a panel made of beechwood showed it to reach equilibrium at 90% r.h. from 45% r.h. in about 20 days for timber ½" thick, 40 days for 1" thick, and 100 days for 2" thick (*Figure 13.5*)[6]. Sudden dry spells can cause cracking and splitting of thick timber.

Fabrics are more complex in their behaviour. The individual fibres swell significantly with increasing moisture and respond in a matter of minutes to changes in ambient moisture. This characteristic for wool is illustrated in *Figure 13.6*[7]. However, on some ropes or multiple-ply threads the result of a swelling of the individual fibres is to tighten the twist, which shortens the overall length. New fabrics can contract initially when exposed to high ambient relative humidities for the first time. This is attributed to a release of the tensions introduced during manufacture and will not recur[3].

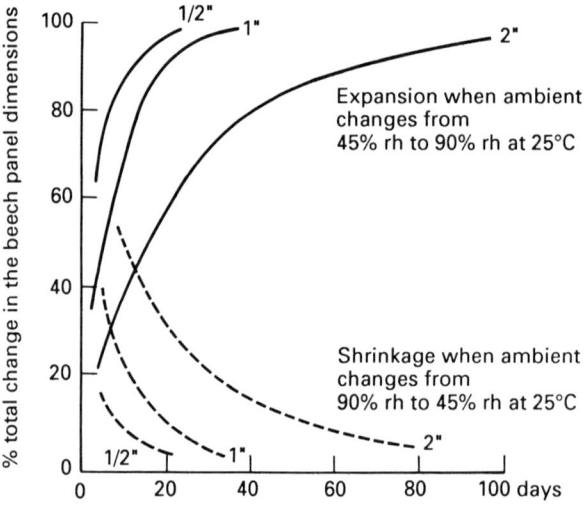

Figure 13.5 Time taken for the expansion and shrinkage of panels of beechwood (Stevens, 1961)[6]

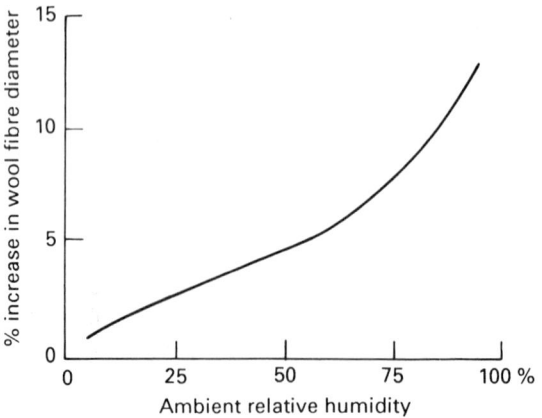

Figure 13.6 The influence of relative humidity on wool fibre diameter (Alexander and Hudson, 1954)[7]

A particular fabric problem is that of the canvas base of oil paintings. Canvas is woven from spun threads of cotton, linen or hemp and in its unprepared state changes its dimensions readily according to the ambient humidity, tightening when damp and expanding when dry. There is also a directional influence with different changes in dimension between the warp and the weft of the canvas[5]. To minimize this influence the canvas is usually prepared by the artist with a coat of glue size. The picture is then built up with layers of different colours. The outer surface is often varnished for preservation. The paint medium shrinks and slowly becomes more brittle over the years. Increases in ambient humidity tend to expand the organic constituents but may lead to a shrinking of the canvas and a distortion of the wooden frame. A prepared canvas of cotton or linen changes its dimensions by 0.05–0.1% for a 10% increase in relative humidity[5]. The combination of these influences can lead to a crazing or cleavage of the paint. Stability of the ambient relative humidity is the most important factor in preservation.

Oil paintings also have a tendency to develop patches of bloom during spells of damp weather. This bloom appears as a bluish film in patches on the surface of varnished oil paintings. It obscures the quality and detail of the picture and is particularly noticeable on dark pictures. The bloom tends to remain even when the air becomes dry again. The bloom appears to be linked to the type of varnish used. Atmospheric pollution, which deposits ammonium sulphate crystals onto the surface which deliquesce whenever the humidity is high, is one cause. The combination of cleaner city air and air conditioning should reduce the incidence[8].

The repair work needed on oil paintings was immediately reduced to almost one-tenth when the paintings were moved temporarily from an old, heated-only building, The National Gallery, London, to an air conditioned depository. As the paintings remained there the repair work decreased and became almost unnecessary[9].

Painted wooden panels exhibit similar characteristics, particularly in large displays made of several individual planks of wood. The changes in moisture affect the dimension across the grain and can in changeable environments lead to shrinkage and cracks in the painting. Experience has shown that most care is needed for the thin wooden panels which are painted on one side. These are common in church furniture. They distort and warp badly in changes of humidity. Thin panels which are painted on both sides have proved to be much more durable[9]. Stability of the ambient relative humidity is essential for preservation here too.

The mechanical properties of materials such as paper and fabric are also significantly affected by their moisture content. When very dry organic fibres become brittle and break easily, glues become inflexible and can break away from their binding surfaces. Increasing moisture contents tend to give a more pliant behaviour. Books are a particularly good example. Very dry conditions can cause embrittlement of the binding adhesive and weakness in the paper and cover hinges which can lead to rapid disintegration of the book when used. The tearing resistance of paper improves rapidly with increasing moisture content. This relationship is illustrated in *Figure 13.7*[10]. The effect of increased moisture is even more dramatic in improving the folding endurance of paper. Increasing the

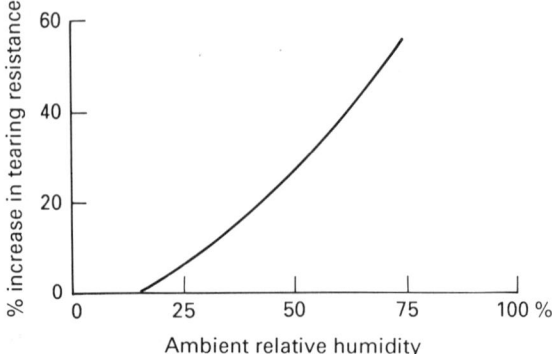

Figure 13.7 Increase in tearing resistance with increasing relative humidity (Carson, 1944)[10]

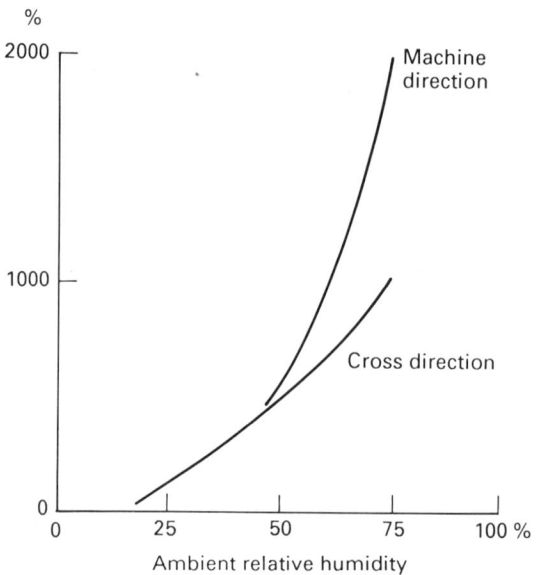

Figure 13.8 Increase in folding endurance of paper with increasing relative humidity (Carson, 1944)[10]

ambient relative humidity from 15% to 65% can increase the folding endurance by 1000% (*Figure 13.8*). Parchment also cockles below 40% r.h., and while this is recoverable by slowly returning to higher humidities, permanent damage may be done to inks and colours on the parchment[11].

The critical factor in avoiding this type of weakness is to avoid too low a relative humidity. There is no consensus on this value but all advisers agree that the ambient relative humidity for textiles and books should be more than 40% and preferably above 45%[3, 12]. More conservative advisers suggest that in practice a design minimum should be 50% r.h.[9, 13] and while no great harm would result for short periods at 45%, prolonged exposure to atmosphere at 45% r.h. may be dangerous to the exhibits.

While the physical conditions have concentrated on relative humidity as the environmental variable, there is a small effect of temperature. This will vary slightly for each material. Wood, illustrated in *Figure 13.2*, typifies the effect of temperature. Dimensional stability is influenced by moisture content and therefore, because the moisture content varies slightly with temperature, this should be considered if very fine precision is needed. While this is not normally the case, the weak interaction between moisture content and temperature and relative humidity is illustrated in *Figure 13.9*. Changes in temperature can be compensated by small changes in relative humidity to keep the moisture content constant. Lowering the temperature from 20°C to 10°C would require the relative humidity to be lowered by 1.5%.

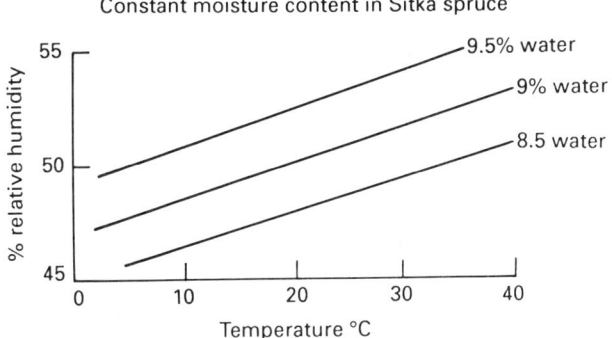

Figure 13.9 Small changes in relative humidity are desirable if moisture content is to be kept constant over a wide temperature range

13.3 Chemical deterioration

13.3.1 Metals

The deterioration route of most metals is corrosion, and the corrosion rate is determined principally by the atmospheric moisture aided by the concentration and type of contaminants. The corrosion rate increases slowly with increasing relative humidity and temperature until a critical relative humidity is reached. Humidities above this critical value have a very pronounced effect on corrosion rate. The important characteristic is therefore this critical relative humidity. It is a function of the metal itself and its impurities, the composition of the ambient air and the presence of any contaminant on the metal surface.

Bronze disease is a powdery, crumbly green mineral, basic cupric chloride, which is formed from cuprous chloride in the presence of atmospheric oxygen. The stability of the patina can be achieved either by removing the cuprous chloride, by protecting it from oxygen or by maintaining low relative humidities. Experiments show that perceptible quantities of basic cupric chloride will form in two hours at 95% r.h. At 35% r.h. the chloride appears to be stable indefinitely. Bronze items therefore benefit from an atmosphere of 35% r.h.[14].

Unstable iron with traces of chloride should be kept below 45% r.h. Good-quality iron, low in sulphur, and bronzes with stable patinas can be kept at 55% r.h. Silver tarnishes faster at high relative humidities. Lead, tin and pewter do not need low relative humidities, and gold is inert to moisture influences[3].

13.3.2 Minerals and glasses

Geological museums note that iron pyrites minerals, either as crystals or pyritized fossils, can oxidize into ferrous sulphate at high relative humidities and therefore, while 55% r.h. is considered the upper limit for relative humidity, the preferred maximum value should be 50%[3].

Some glass objects are also sensitive to high ambient relative humidities. If the proportion of calcium, magnesium and aluminium is low, then over many years traces of sodium and potassium are leached out from the glass in the form of hydroxides. These hydroxides are converted by carbon dioxide in the atmosphere to carbonates, which are deliquescent. At high relative humidities this results in water droplets forming or weeping on the surface[9]. As the deterioration progresses, the glass may develop tiny cracks and eventually become opaque. Similar experiences have been recorded for lenses in telescopes, binoculars and cameras. Accelerated experiments at 90°C showed polished glass sheets to achieve the same visible deterioration after 24 hours of exposure at 90% r.h., 360 hours at 73% r.h. and 1296 hours at 57% r.h. Physical measurements of transmission loss through the glasses did not correspond with their appearance. In practice the storage of such goods in tropical warehouses can lead to marked deterioration in one month[15]. An upper limit of 50% r.h. is proposed to prevent this deterioration[16].

13.3.3 Paper, textiles and leather

One major worry with coloured materials is that in time the colour fades. Careful research on organic dyes painted onto paper has shown how the speed of fading is affected by temperature, ambient relative humidity and ambient gas composition[17]. The results are illustrated in *Figure 13.10*. The work shows that a given light source fades the colours much more rapidly at high ambient relative humidities. Increasing the temperature from 20°C to 30°C almost doubled the rate of fading. The magnitude of the effect was similar for a range of popular organic dyes. The curators of fabrics accept that this will apply in general to textiles but are more concerned about the more important embrittlement and resulting damage which low relative humidities have on textiles[18].

The experiments also showed how a nitrogen-filled atmosphere reduced the speed of fading to approximately one-tenth (1/6th to 1/19th) of the speed in air.

Light also has a weakening effect upon paper. Experiments showed that this effect was modified by humidity. High humidities accelerated the deteriorating effect of light on paper made from cotton cellulose but moisture had the reverse effect on paper made from wood pulp[19].

Atmospheric air often contains traces of acid gases, particularly sulphur dioxide. When this gas comes into contact with water it forms sulphurous

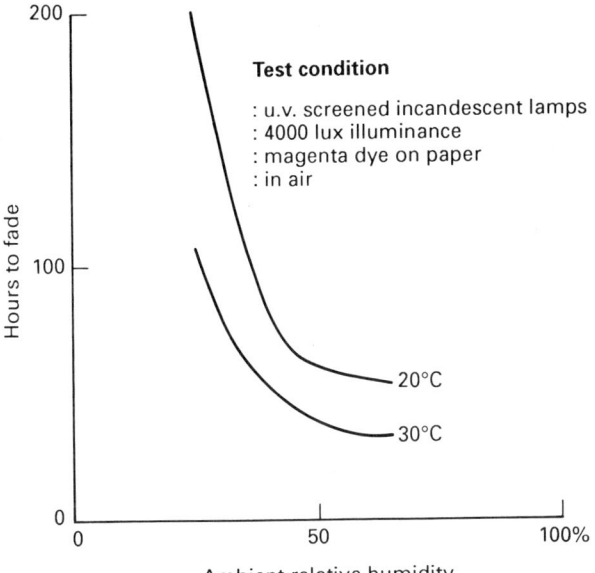

Figure 13.10 The influence of relative humidity on time to fade an organic dye used in water colour painting (Kuhn, 1968)[17]

acid, which, although a weak acid, can over time destroy organic materials such as paper. Experiments have shown that this attack rate is much higher at higher relative humidities and at higher temperatures[20].

13.4 Biodeterioration

The two major biodeterioration routes are from microorganisms, such as mould, and insects such as the clothes moth. In general, microorganisms, particularly mould, which creates mildew and the musty smell, are more of a museum problem in warm, humid climates, while insect damage is more troublesome in the more temperature climates. Let us look at the factors which influence the risk of damage.

13.4.1 Microbiological spoilage

Mould spores are present in everyday air. They require moist conditions before they can germinate, and nutrients before they can grow. Most organic materials in a museum are a suitable source of nutrients[21].

The germination and growth of mould spores is linked to temperature and moisture. There is a critical relative humidity below which mould spores remain dormant. Above that critical value the speed of growth is directly linked to increasing relative humidity, although at very high moisture levels bacteria tend to develop and take over from the moulds. Each mould has an optimum growth temperature and its tolerance to moisture is highest at this optimum temperature. An illustration of the

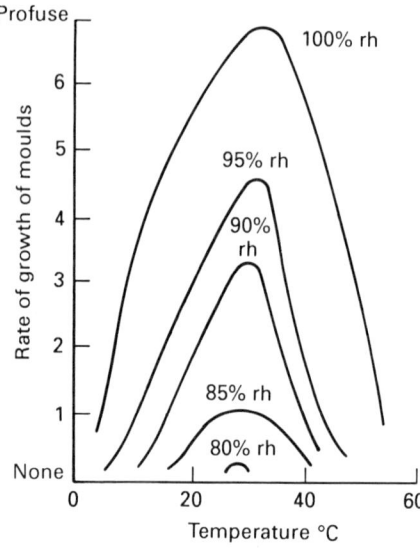

Figure 13.11 The growth rate of a mixture of moulds on textile and packaging materials (Burns, 1925)[22]

relationship between mould growth and temperature and humidity is given in *Figure 13.11*[22]. This is for a mixture of moulds on textile and packaging material. Each mould has its own characteristic relationship with temperature and humidity.

The critical relative humidity is that in the immediate vicinity of the mould spore. If the mould spore is resting on the inner surface of an outer wall of the museum then the inner surface of this outer wall will be cooler than the ambient air within the room. This means that for the same water moisture content in the air, the relative humidity in the air adjacent to the wall will be higher. This is illustrated in *Figure 13.12*. The temperature difference will be small for the modern, well-insulated buildings but could be 5°C for old constructions. This means that the relative humidity could be 20% more at the wall surface.

A similar effect is achieved if unheated and closed cupboards or displays are fixed to an outer wall. The relative humidity inside the cabinet can become unacceptably high, and even though the temperature is low, mildew may occur in time. This problem has already been reported in museums, and the cure is to move the cabinet to an internal wall of the museum[23].

The same principle also applies to boxed exhibits being transported from place to place and undergoing a large change in temperature. A well-sealed packing case loaded in a warm, humid climate becomes very damp if stored or carried in cold conditions. One solution is to introduce a moisture buffer. In more recent times the advantages of using deliberately introduced moisture buffers have been recognized. Silica gel, conditioned to the required relative humidity, has an outstanding moisture-absorbing capability with little hysteresis in its adsorption/desorption cycle[1,24,25].

The damage effected by microorganisms varies from surface blooming with little permanent damage, through discoloration, to complete disintegration of the material. The susceptibility of leather varies with its

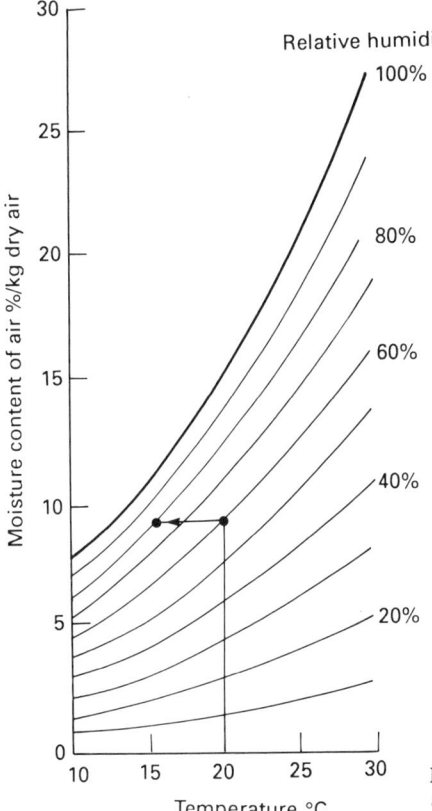

Figure 13.12 The relative humidity rises as air cools

pre-treatment. Chrome-cured leather is not hygroscopic and is therefore particularly durable and resistant to attack by mould. Vegetable-cured leathers are very susceptible but the damage is negligible and the bloom is readily removed. Cellulosic materials such as paper, cotton, jute, sisal, hemp and linen are more susceptible to mildew than the keratinous materials such as wool, felt and fur. Dirt on textiles increases their susceptibility to attack because of the enhanced nutrients. Discoloration in fabrics due to moulds tend to be diamond-shaped, as microbiological growth occurs from the germination focal point along the warp and weft fibres. The discoloration of the fabric is usually permanent because the acid metabolites of the fungal growth process can chemically react with the dyes. Canvas painted with chrome green often becomes spotted with yellow patches. Ultramarine tints can become decolorized and cloth dyed with aniline can turn green. Fabric which has been attacked loses its strength and tears or punctures on handling[26]. Books are also prone to mould attack because of the particularly rich supply of nutrient present in the gelatine, the size glue and the bookbinding paste which are used in the bookbinding process.

Experiments on the ease with which mould grew on different types of paint binder showed large differences when the materials were exposed to relative humidities between 68% and 98%. The degree of resistance of the popular binders is illustrated in *Figure 13.13*[27].

Fungal attack on timber is rare in the absence of a building defect or failure which has accidentally permitted the timber to become saturated with water. Timber below 20% moisture content is considered safe from such damage. This is equivalent to an equilibrium relative humidity of approximately 90%[2].

In practice, fungal deterioration can be avoided by keeping the relative humidity below 68% and preferably below 65% at normal temperatures of 15–25°C[9].

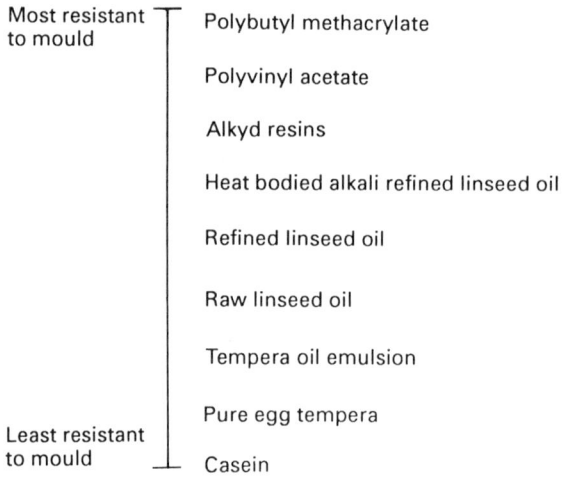

Figure 13.13 Recommended bands of relative humidity for different materials

13.4.2 Insect damage

In temperate climates insects cause more damage than moulds. Wool and similar keratinous materials such as fur and felt are the most susceptible. This means that textiles deserve special attention. Surveys show that almost 80% of textile damage can be attributed to moths or carpet beetles in equal share[28, 29].

Unfortunately, while insects have optimum relative humidities which are generally high (the clothes moth *Tineola* favours 65–75% r.h., the house moth *Hofmannophila* about 90% r.h.), insects are relatively tolerant to a wide range of humidities. Very low relative humidities, about 25% r.h., will retard the growth of moths and carpet beetles but not eliminate them[29].

Wood-boring insects can also damage books which stand in their way. One, the power post beetle (*Lyctus*) only develops in hardwood and is sensitive to the moisture content of the wood. The eggs and larvae can develop in moisture contents from 10 to 28% but not below 8%, which is approximately in equilibrium with a relative humidity of about 50%[2].

Woodworm is the worst British insect pest and favours dry timber. Regular checks of furniture for tiny worm holes, including any plywood backing panels, is essential.

Moisture control is therefore not a suitable technique for preventing insect damage.

13.5 Conclusions

(1) The control of relative humidity in museums is important. Since people cannot readily sense changes in humidity, some form of instrument sensor is required.

(2) Relative humidity is important in two different ways. The first is the value of the relative humidity and the second is the stability of this value. For thin materials which respond rapidly to moisture changes, constancy of relative humidity could be more important than the actual value chosen. This is particularly true for materials protected on one side but exposed to moisture on the other. Warping, distortion and cracking can occur.

(3) The best values of relative humidity vary widely with the exhibit material. Very dry conditions inhibit corrosion of metal objects but can

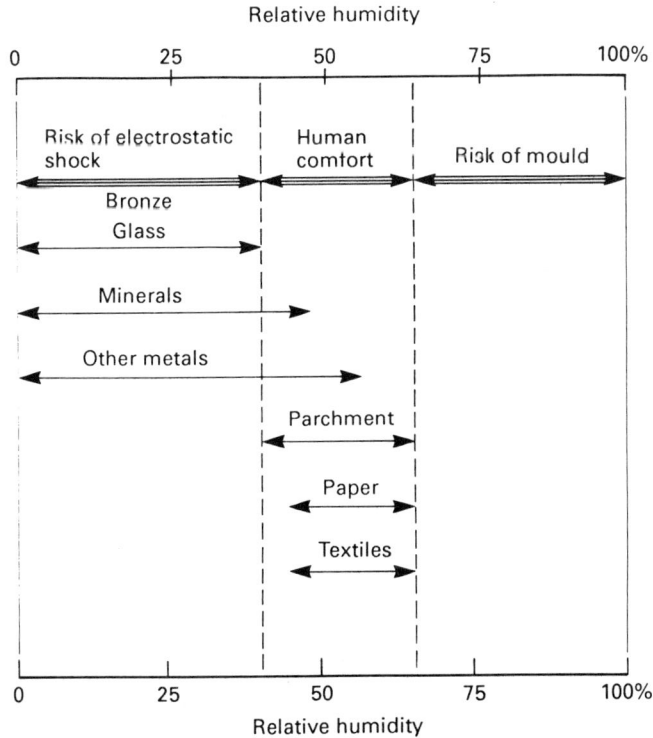

Figure 13.14 Recommended humidity for museum collections

cause embrittlement in organic material such as parchment, fabrics and books. A summary of recommended values is given in *Figure 13.4*.

(4) Controlled air conditioning has been demonstrated to be the most satisfactory way of preserving exhibits. Economical ways of operation are being explored, particularly with a view to slowly changing the design conditions over the year[30].

(5) Simple techniques are now in use to minimize the wide changes in relative humidity which can occur. In temporary transit this is now achieved by introducing a buffer material such as conditioned silica gel into the transit container. In the slower, more seasonal changes of moisture, the concept of a protected room sited within another room offers some stability of temperature and protection from excess ventilation[1, 31, 32].

(6) Historic houses which are open to the public during the summer are now becoming moisture controlled through dehumidification when they are closed during the winter, rather than by the much more expensive route of space heating. Design values of 50–65% r.h. are the best compromise for cost and acceptability[33].

References

1 Toishi, K. 'Humidity control in a closed package'. *Studies in Conservation*, **4**, 81–87, 1959
2 Kollman, F.F.P. and Cote, W.A. *Principles of Wood Science and Technology*, Vol. I. *Solid Wood*. Springer-Verlag, New York, 1968
3 Thomson, C. *The Museum Environment*. Butterworths, London, 1978
4 Buck, R.D. 'A note on the effect of age on the hygroscopic behaviour of wood'. *Studies in Conservation*, **1**, 39–44, 1952
5 Cornelius, F. du Pont. 'Movement of wood and canvas for paintings in response to high and low r.h. cycles'. *Studies in Conservation*, **12**, 76–80, 1967
6 Stevens, W.C. 'Rates of change in the dimensions and moisture content of wooden panels resulting from changes in the ambient air conditions'. *Studies in Conservation*, **6**, 21–25, 1961
7 Alexander, P. and Hudson, R.F. *Wool: its Chemistry and Physics*. Chapman and Hall Ltd, London, 1954
8 Brommelle, N. 'Bloom in varnished paintings'. *Museum Journal*, **55**, 263–266, 1956
9 Plenderleith, H.J. and Werner, A.E.A. *The Conservation of Antiquities and Works of Art*. Oxford University Press, London, 1971
10 Carson, F.T. 'Effect of humidity on physical properties of paper'. NBS Circular C445. Washington, USA, 1944
11 Faraday, M. *On the Ventilation of Gas Burners*, 1843. Quoted in *Progress in Leather Science 1920–1945*. British Leather Manufacturers Research Association, Egham, 1946
12 Leene, J.E. (ed.). *Textile Conservation*. Butterworths, London, 1972
13 Wener, A.E.A. 'Heating and ventilation'. *Museums Journal*, **57**, 159–166, 1957
14 Organ, R.M. 'The examination and treatment of bronze antiquities'. *Recent Advances in Conservation: Contributions to the IIC Rome Conference 1961* (ed. G. Thomson) pp. 104–110. Butterworths, London, 1963
15 Stockdale, G.F. and Tooley, F.V. *J Am Chem Soc*, **33**, 1116, 1950
16 Werner, A.E. 'Care of glass in museums'. *Museum News*, Technical Supplement, 45–49, June 1966
17 Kuhn, H. 'The effect of oxygen, relative humidity and temperature on the fading rate of water colours. Reduced light damage in a nitrogen atmosphere'. London Conference on Museum Climatology 1967: Rev. Edition IIC 79–88, 1968
18 Thomson, C. 'Textiles in the museum environment'. Chapter 7, pp. 98–112 in *Textile Conservation*, (ed. J.E. Leene). Butterworths, London, 1972

19 Launer, H.F. and Wilson, W.K. *J Res Natl Bur Standards,* **30**, 55–74, 1943. Wessel, C.J. 'Paper', p.389 in Chapter 6, pp.357–407 in *Deterioration of Materials* (eds G.A. Greathouse and C.J. Wessel). Reinhold Pub. Corp., New York, 1954

20 Kimberley, A.E. and Scribner, B.W. 'Summary report of National Bureau of Standards research on preservation of records'. US National Bureau of Standards Miscellaneous Publication M154, March 1937

21 Plenderleith, H.J. 'Mould in the muniment room'. *J Archives,* **7**, 13–18, 1952

22 Burns, A.C. 'Mould growth on textiles and packaging material'. *J Text Inst,* **16**, T185–191, 1925

23 Buck, A. and Leene, J.A. 'Storage and display'. Chapter 8, pp. 113–127 in *Textile Conservation* (ed. J.E. Leene). Butterworths, London, 1972

24 Thomson, G. 'Relative humidity – variation with temperature in a case containing wood'. *Studies in Conservation,* **9**, 153–169, 1964

25 Thomson, G. 'Stabilisation of r.h. in exhibition cases: hygrometric half time'. *Studies in Conservation,* **22**, 85–102, 1977

26 Siu, R.G.H. *Microbial Decomposition of Cellulose.* Reinhold Publishing, New York, 1951

27 Boustead, W. 'The conservation of works of art in tropical and sub-tropical zones', pp. 73–79 in *Recent Advances in Conservation: Contributions to the IIC Rome Conference 1961.* Butterworths, London, 1963

28 Laibach, E. 'Insekten als Schadlinge an Textilien'. *Z ungew Ent,* **47**, 142–148, 1960

29 Hueck, H.J. 'Textile pests and their control'. Chapter 6, pp. 76–97 in *Textile Conservation* (ed. J.E. Leene). Butterworths, London, 1972

30 Stolow, N. 'Conservation of exhibits in the museums of the future', pp. 24–31 in *Conference Proc for 2001 – The Museum and the Canadian Public.* Canadian Museums Association, 1977

31 Stolow, N. 'Fundamental case design for humidity sensitive museum collections'. *Museum News,* Technical Supplement, **11**, 45–52, February 1966

32 Padfield, T. 'The control of relative humidity and air pollution in show cases and picture frames'. *Stud Conserv,* **11**, 8–30, 1966

33 Staniforth, S. and Hayes, B. 'Temperature and relative humidity measurements and control in National Trust Houses'. *ICOM Committee for Conservation Working Group 17, Lighting and Climate Control,* Vol. 3, pp. 915–926, 1987

Further reading

Amdar, E.J. 'Humidity control – isolated area plan'. *Museum News Tech Supplement,* **5**, 58–60, December 1964

Block, S.S., Rodriquez-Torrent, R., Cole, J.B. and Prince, A.E. 'Humidity and temperature requirements of selected fungi'. *Dev Ind Microbiol,* **33**, 204–216, 1962

Evans, D.M. 'The deterioration of bookbinding materials', pp. 179–184 in *Microbiological Deterioration in the Tropics.* Society of Chemical Industry Monograph 23, London, 1966

Greathouse, G.A. and Wessels, C.J. *Deterioration of Materials.* Reinhold Publishing Corp., New York, 1954

Hearle, J.W.S. and Peters, R.H. *Moisture in Textiles.* Textile Institute and Butterworths Scientific Publications, Manchester and London, 1960

Hinton, H.E. *A Monograph of the Beetles Associated with Stored Products.* British Museum (Natural History), London, 1945

McKenny Hughes, A.W. 'Insect pests of books and paper'. *Archives,* **7**, 19–22, 1952

MacLeod, K.J. *Relative Humidity: its Importance, Measurement and Control in Museums.* Canadian Conservation Institute Technical Bulletin No. 1, 1975, Ottawa

Mallis, A. *Handbook of Pest Control.* MacNair Dorland, New York, 1960

Metcalf, C.L. and Flint, W.P. *Destructive and Useful Insects, their Habits and Control.* McGraw Hill, New York, 1930

Langwell, W.H. (ed.). *The Conservation of Books and Documents.* Pitman, London, 1957

Law, P.G. 'The problem of condensation in optical instruments'. Australian Scientific Liaison Office, Washington DC, USA, Australian Technical Paper 990, August 1945

Lehmann, D. 'Conservation of textiles at the West Berlin State Museum'. *Stud Conserv,* **9**, 15, 1964

Plenderleith, H.J. and Werner, A.E.A. *The Conservation of Antiquities and Works of Art*, 2nd edition. Oxford University Press, London, 1971

Rousseau, M.Z. 'Sources of moisture and its migration through the building enclosure'. *ASTM Standardisation News*, **12**, 35–37, November 1984

Thomson, G. (ed.). *Recent Advances in Conservation*. Butterworths, London, 1963

Weiss, H.B. and Carruthers, R.H. *Insect Enemies of Books*. New York Public Library, 1937

Whitfield, S., Cole, F.G. and Whitney, G.F.H. 'The bionomics of *Toneola bisselliella* Humm under laboratory culture and its behaviour in biological assay'. *Lab Pract*, **7**, 210–217, 275–284, 339–343, 408–411, 1958

Chapter 14

The measurement of moisture in gases

14.1 Introduction

The moisture content of gases is one of the most difficult environmental variables to measure accurately and reliably. The measuring instruments are also amongst the most difficult to calibrate and the manufacturers' claims amongst the most optimistic. However, the measurement technology is now one of the most rapidly developing. This is attributed to two factors. The first is the exploding market demand for moisture sensors. They are about to become a key element in the carburation control of automobiles and lorries to aid optimum economy. They are already being incorporated into microwave cookers for added convenience and into mechanical ventilation systems in houses to provide condensation control.

There are five major types of moisture sensor:

(1) *Measurements of physical changes*. This the most widespread type of hygrometer and the sensor usually measures the change in length of hair or a nylon strip with changes in relative humidity. Calibration is essential.

(2) *Psychrometry*. This utilizes a knowledge of the heat and mass transfer in an air stream to convert the readings of a wet bulb thermometer and a dry bulb thermometer into moisture content of the ambient air. This is a low-cost reliable technique and is usually used for spot-checks. No calibration is needed.

(3) *Salt hygrometer*. This uses the physical properties of saturated salt solutions, such as lithium chloride, to determine the dew point of the ambient air. Such sensors are now physically small and are widely used for continuous monitoring of gases.

(4) *Peltier dew point*. This uses the Peltier solid-state refrigeration technique to chill a mirror down to dew point. It is widely used for low-temperature applications and can incorporate advanced electronic temperature controls.

(5) *Changes in electrical properties*. When hygroscopic materials adsorb water vapour their electrical properties of both resistance and capacitance

change. New solid-state permeable bulk sensors are now entering the market with the expectation that they will be more reliable and less liable to contamination than some of the surface-coated sensors. This is an area of rapid change.

Let us consider these five in more detail.

14.2 Measurement of physical changes with moisture

Most hygroscopic materials change their dimensions with the amount of moisture adsorbed or absorbed. Their equilibrium moisture content is a function of the ambient relative humidity with a minor modification for temperature. The dimensional changes of such materials can therefore be used to indicate the ambient relative humidity. Three materials are in widespread commercial use: hair, nylon and gold beaters skin.

(a) *Hair sensor*
Animal hair, suitably cleaned and degreased, has been used for well over 100 years as a humidity sensor (*Figure 14.1*)[1]. Its change in length with changes in relative humidity are illustrated in *Figure 14.2*. The relationship is almost linear. Hysteresis is small. The change in length with temperature ($19 \times 10^{-6}/°C$) is very similar to that of brass, and therefore in a brass holder the hair sensor is relatively insensitive to the influence of ambient temperature. The response time to changes in relative humidity is of the order of a few minutes (*Figure 14.3*)[2] and is independent of temperature. Rolling the hair to give it an ellipsoidal cross-section gives a much faster response time but at the expense of considerably weakened mechanical strength.

The effect of dirt can increase the readings by 10 percentage points at high relative humidities. One illustration of the effect of cleaning an 18-year-old hair sensor with diethyl ether is shown in *Figure 14.4*. This is an extreme example. The sensor is not very sensitive to handling, providing the hands are clean.

The sensor is recommended for applications above 30% r.h. and above $-10°C$.

(b) *Nylon sensor*
Nylon 6 has replaced hair in the more compact sensors. It has about three times the change in length with humidity compared to that for hair (*Figure 14.2*). It has little hysteresis and is more responsive to changes in relative humidity than hair at temperatures down to 10°C. Its response to changes in relative humidity below 10°C becomes very slow and can take five hours at 5°C.

The sensor is recommended for temperatures above 10°C.

(c) *Gold beaters skin*
This is prepared from the outer membrane of the large intestine of the ox. It was widely used in separating sheets of the fine gold used by decorators. Its physical dimensions are similar to those of the nylon strip, typically

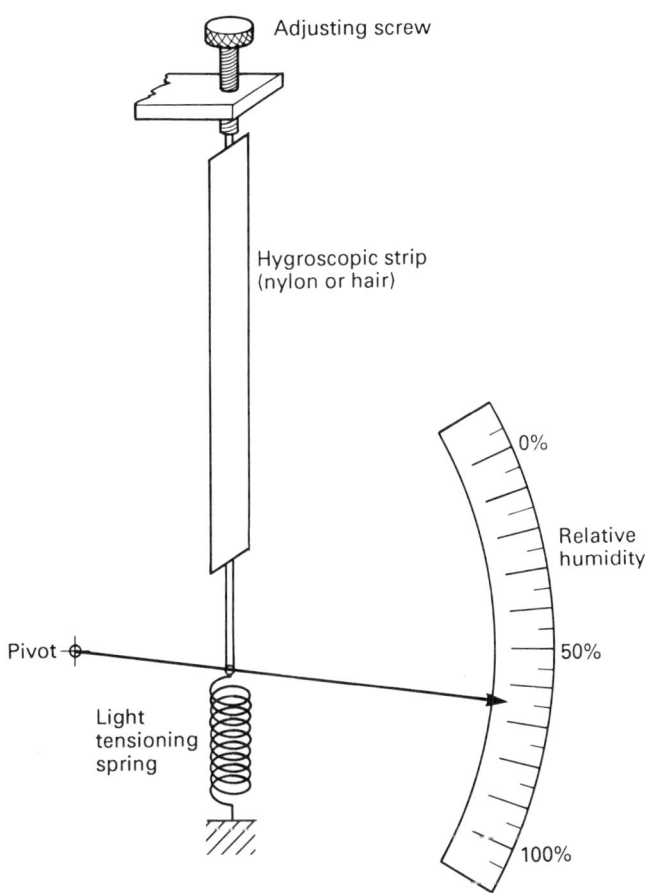

Figure 14.1 An illustration of an extension hygrometer

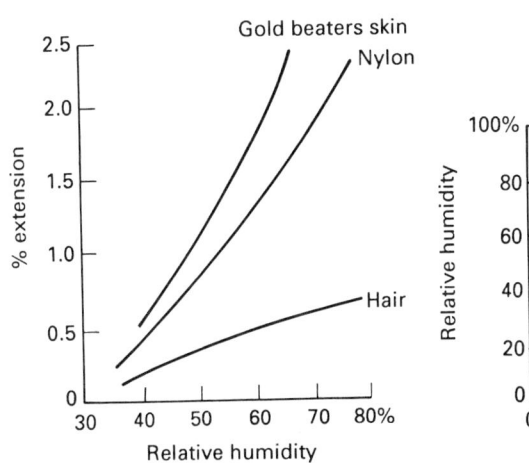

Figure 14.2 Extension characteristics of some humidity sensors at 35°C

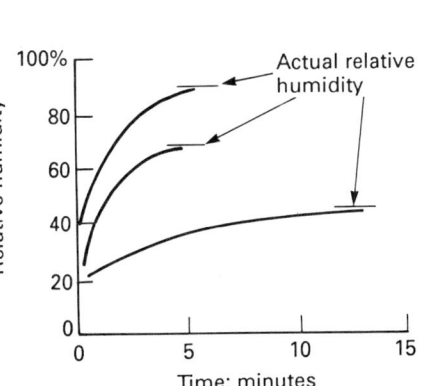

Figure 14.3 Illustrative response times for a hair hygrometer

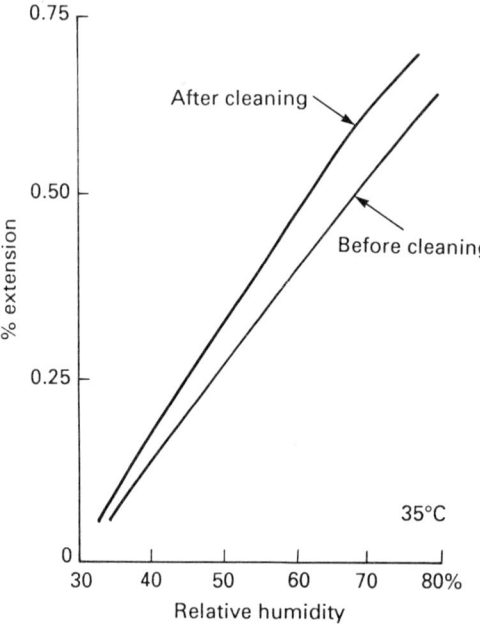

Figure 14.4 The effect of cleaning a hair humidity sensor

200 mm × 25 mm × 0.02 mm. Its change in length with changes in relative humidity are greater than those of nylon but there is a large hysteresis at low relative humidity. An illustration of this effect is shown in *Figure 14.5*.

The mass of humidity sensors also changes with ambient relative humidity. This property is used in a piezo-electric hygrometer. A

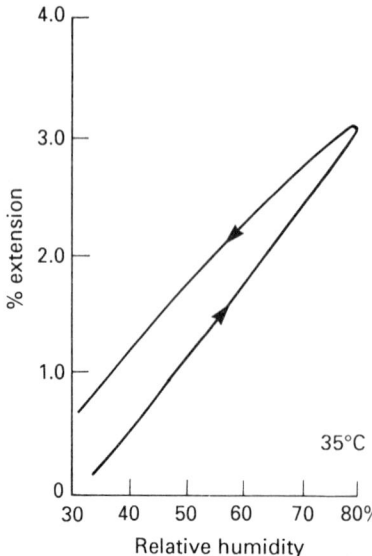

Figure 14.5 Effect of hysteresis on a gold beaters skin humidity sensor

piezo-electric crystal is a substance, usually synthetic quartz, which changes shape when placed in an electric field. A thin sheet of this quartz, with electrodes on both surfaces, will vibrate if mounted in an electric oscillator circuit. The presence of mass loading on the crystal surface affects the oscillation frequency inversely and proportionally to the mass. Thus a hygroscopic coating which will reversibly absorb moisture from the surrounding atmosphere can be weighed by recording the decrease in oscillation frequency[3].

Infrared absorption can also be used to measure humidity. The absorption spectrum for water vapour is very selective at certain wavelengths and these wavelengths can be used to quantify the amount of moisture in the beam path. Typical beam paths are 0.25 m and the infrared source is collimated and chopped by a rotating filter wheel transmitting alternately at 2.6 μm and 2.45 μm. A lead sulphide detector records the transmitted energy. The 2.6-μm beam is particularly attenuated by the moisture in the beam while the slightly shorter wavelength is relatively unaffected. The attenuation of the longer wavelength relative to the shorter is therefore proportional to the water vapour content in the infrared beam[4].

14.3 Psychrometry: wet and dry bulb hygrometry

In a saturated atmosphere no evaporation occurs and therefore the temperatures recorded by two thermometers, one with a wet bulb and the other with a dry bulb, would be the same. If the atmosphere was not saturated then some evaporation would occur from the wet bulb thermometer, cooling it. The temperature depression created will increase as the ambient relative humidity falls. Knowledge of the dry bulb and wet bulb temperature allows the ambient relative humidity to be calculated.

The apparatus is illustrated in *Figure 14.6*. The temperature depression is very sensitive to the air stream velocity. This effect is shown in *Figure 14.7*. In practice it is more convenient to arrange a planned air flow over the thermometers at around 3 m/s to avoid unplanned local velocities influencing the results, although the recent World Meteorological Organisation publication requests 4–11 m/s[5]. The wick length must be at least 100 mm[6].

Accuracy is ±3% r.h., providing the thermometers themselves are accurate and the wet bulb wick clean. The capital cost is low and the instrument is reliable. This makes it the most popular method of recording humidity.

14.4 Salt hygrometry

In general, the two factors influencing the water vapour pressure of a salt solution are the salt concentration and the temperature. Increasing the salt concentration lowers the vapour pressure, while increasing the tempera-ture increases the vapour pressure. If the salt solution remains saturated with an excess of salt present, then for each dew point or vapour pressure

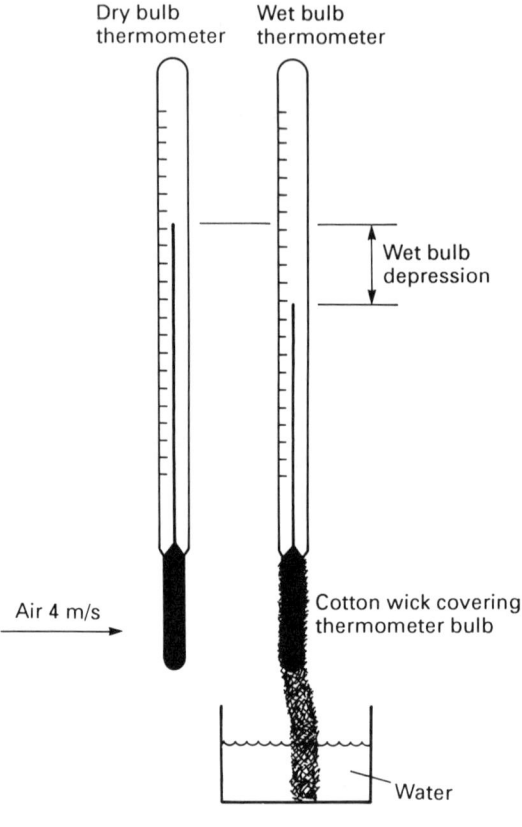

Dry bulb
thermometer

Wet bulb
thermometer

Wet bulb
depression

Air 4 m/s

Cotton wick covering
thermometer bulb

Water

Figure 14.6 Illustration of a wet
and dry bulb hygrometer

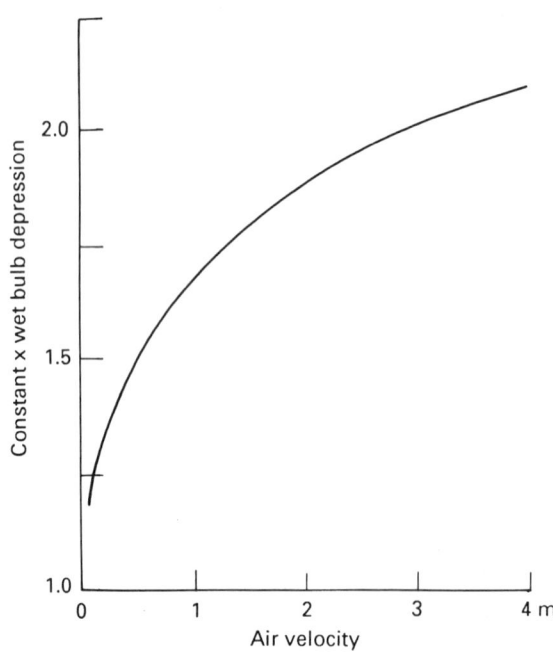

Figure 14.7 The effect of air
velocity on wet bulb
depression

of the water vapour in the air, there will be a unique temperature of the saturated salt solution which will be in equilibrium with it. This relationship for a saturated lithium chloride solution is illustrated in *Figure 14.8*[7].

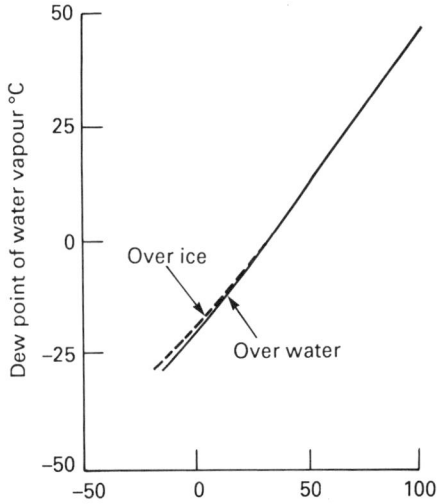

Figure 14.8 Relationship between the dew point of saturated lithium chloride and that of water vapour

Reliable hygrometers are based on this relationship. Their design is influenced by the electrical properties of a salt. In solution the salt ionizes and permits a ready flow of electricity. As a crystal deposit the salt is almost an electrical insulator. This characteristic can be used to provide a self-heating sensor. The construction of a typical sensor is shown in *Figure 14.9*. An absorbent fabric is wrapped around an insulating holder. This fabric is saturated with a lithium chloride solution. Two electrical leads are

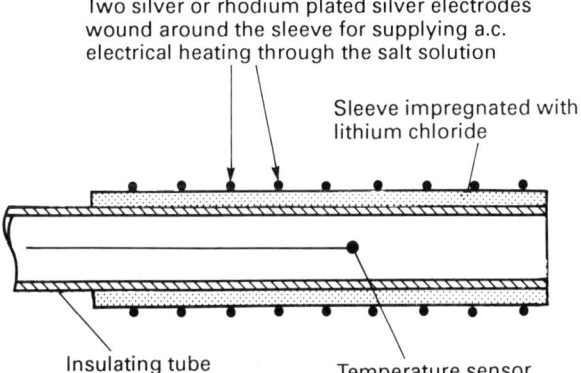

Figure 14.9 Diagrammatic illustration of a lithium chloride dew point sensor

wrapped around the fabric with the electrical path between them being through the lithium chloride impregnated fabric. When the probe is placed in a moist atmosphere, the sensor will at first absorb the moisture and create an electrically conducting path between its electrical leads. This current will heat up the solution until the temperature is such that the salt starts to crystallize out. The electrical power will then decrease and the sensor will reach an equilibrium temperature[8,9]. This equilibrium temperature will be the dew point of this salt solution, and the dew point of the water vapour of the air can be determined from the fundamental characteristics of the salt given in *Figure 14.8*.

The sensor is very reliable but will not work below 11% r.h. at room temperature and must not be used at very high relative humidities, because of the possible washing away of the salts. It is also sensitive to contamination and some models when not in use may corrode. A typical working range is illustrated in *Figure 14.10*[10].

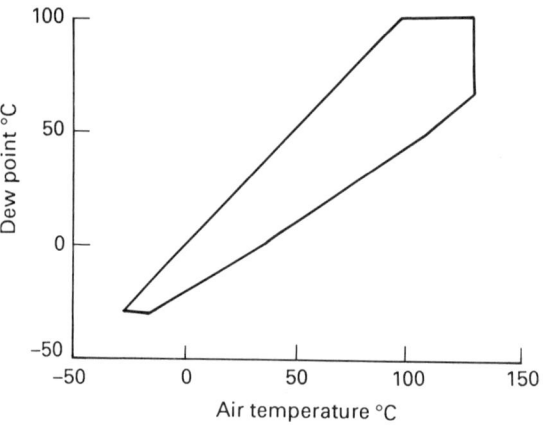

Figure 14.10 Illustrative measuring range for a lithium chloride sensor

14.5 Peltier driven mirror dew point sensor

When a surface is cooled to the dew point of the water vapour in the gas, then condensation occurs and it can be identified as beads of moisture, or frost if the temperature is below freezing point, on the mirror surface. The technique has long been used manually, for example by blowing air through ether to provide local cooling by evaporation. However, the availability of Peltier modules means that a d.c. electric current can be used to pump heat from the mirror surface. Such electrical techniques lend themselves to automatic control and the reflected and scattered light detectors can be used to give a signal to drive the mirror temperature down to the dew point temperature and hold it there with precision. The mirrors are usually rhodium plated silver. The rhodium is used for its hardness and the silver body for good thermal conductivity.

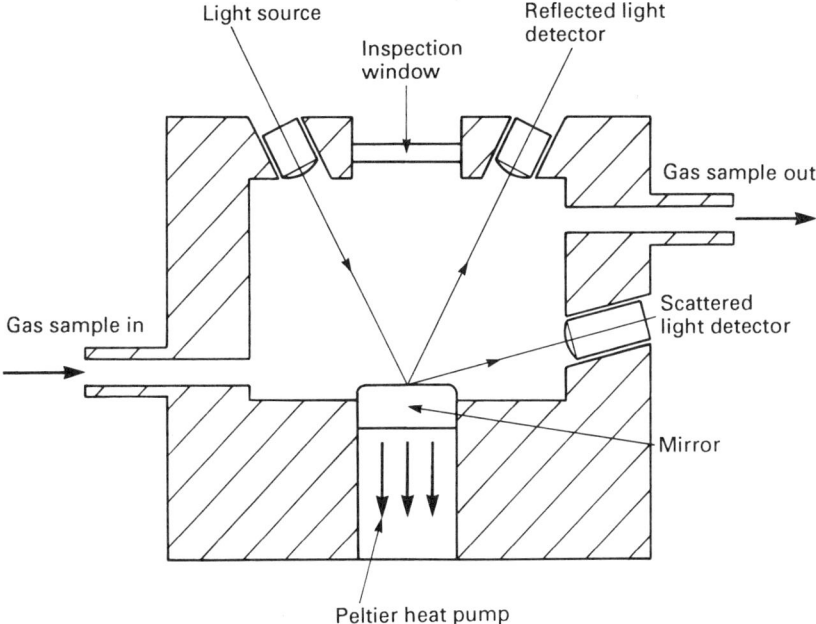

Figure 14.11 Peltier cooled mirror dew point sensor

An illustration of such a cell is given in *Figure 14.11*[11]. No calibration is necessary. The lowest operating temperature is determined by the heat flow from the body of the sensor, but manufacturers claim satisfactory operation down to a dew point of −60°C.

14.6 Electrical characteristics as a function of moisture

There have been many humidity sensors which relied on the surface conducting properties of the sensor element. The surface was treated with a hygroscopic salt such as lithium chloride and the electrical resistance between conducting grids attached to the surface was measured.

Such probes quickly become contaminated or lose their surface treatment if accidentally exposed to very high relative humidities. Sensors now tend to use bulk effects, not surface effects. A bulk sensor minimizes the contamination problems of the surface sensors because the contaminants do not penetrate into the bulk of the sensor.

An illustration of a bulk polymer sensor is given in *Figure 14.12*. In humid conditions mobile ions are released from the quaternary ammonium macromolecular structure in the bulk polymer film. Changes in humidity increase or decrease the ion activity and hence influence the electrical resistance of the bulk of the sensor. One manufacturer, General Eastern, uses a matched diode temperature compensation and provides a 1-kHz electrical supply to avoid polarization. The resistance change with humidity is given in *Figure 14.13*[12].

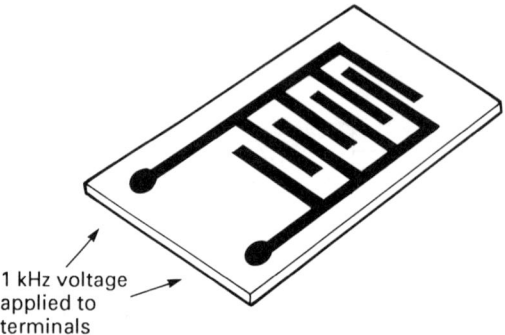

Figure 14.12 Illustration of a bulk polymer resistance sensor

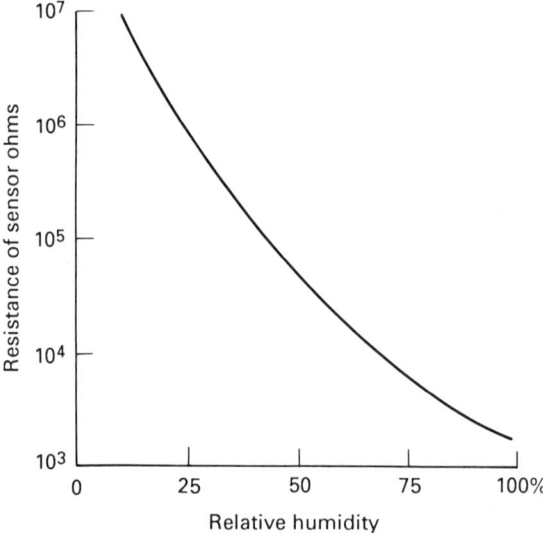

Figure 14.13 The effect of ambient relative humidity on resistance of the bulk polymer sensor

Metal oxide ceramics are now available as bulk humidity sensors. These sensors detect humidity on the basis of the enhancement of the surface electrical conductivity of oxides by water adsorption within the porous sensor. Manufacturers claim greater durability because the ceramic is more stable both physically and thermally. The structure of such a sensor is illustrated in *Figure 14.14*. This is manufactured from the oxides of zinc, chromium, vanadium and lithium. This forms a binary system $ZnCr_2O_4$–$LiZnVO_4$ whose porosity can be decided by the ratio of $ZnCr_2O_4$ and $LiZnVO_4$. The ceramic body is mainly constructed from $ZnCr_2O_4$ spinel grains with the vanadium compound in the glassy phase on the spinel grains[13].

The surfaces of most metal oxides are covered with –OH hydroxyl groups because physically adsorbed water transforms to surface hydroxyl groups. The reliability of this particular oxide is attributed to the glassy

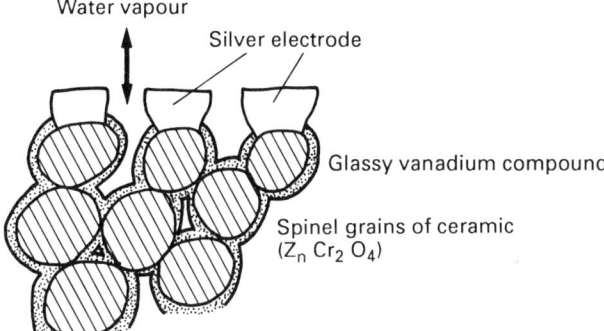

Figure 14.14 Close-up of the microstructure of the ceramic bulk sensor

vanadium complex, which includes water molecules in its gel-like structure, and therefore the water adsorbed onto this surface transforms only slightly to surface hydroxyl groups.

The change in the electrical properties of resistance and capacitance are illustrated in *Figure 14.15*. The time taken to respond to changes in ambient relative humidity is shown in *Figure 14.16*, typically 150 seconds.

The robustness of the ceramic bulk sensors means that they can be heat-treated to remove condensed oils and fats. This is particularly useful in some catering and industrial applications. One example which successfully regenerates the sensor by momentarily heating it up to 500°C is shown in *Figure 14.17*[14].

Microchips 1.5 mm square with both a miniature humidity sensor and a temperature sensor are becoming available. Research has shown that a hygroscopic polymer such as cross-linked cellulose acctate butyrate can be bonded onto a gold–titanium electrode 3000 Å thick. The upper electrode

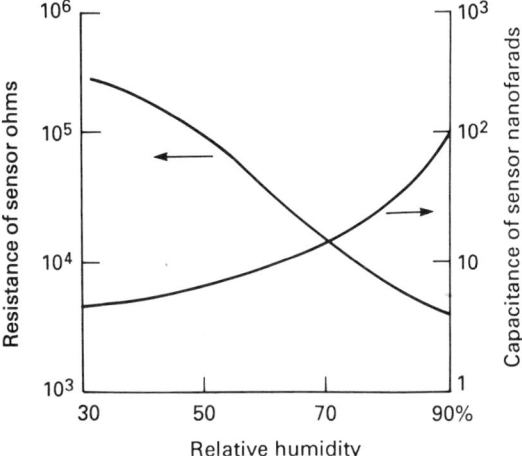

Figure 14.15 Electrical characteristics of a ceramic bulk humidity sensor

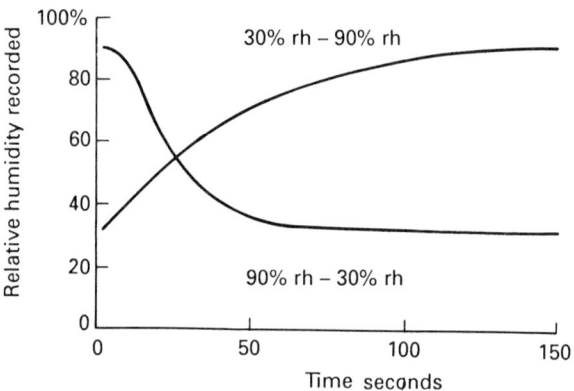

Figure 14.16 Measured response of the ceramic bulk humidity sensor

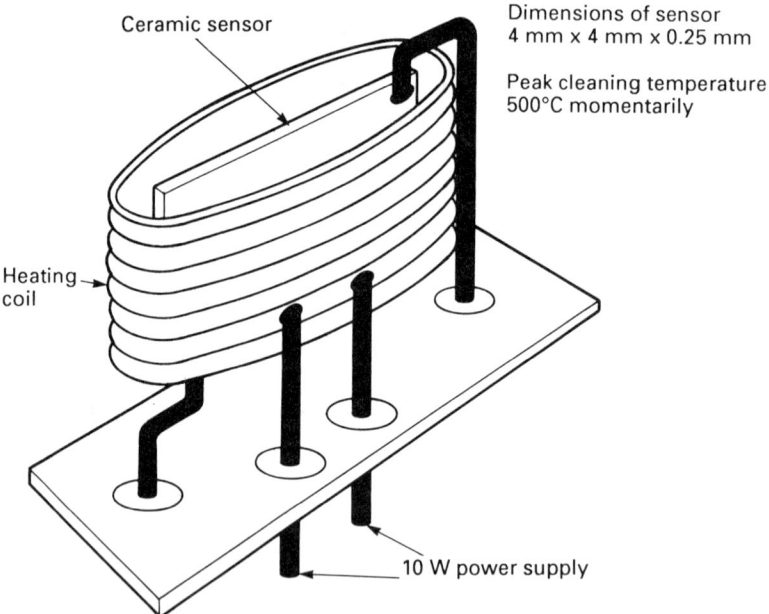

Figure 14.17 Construction of self-cleaning bulk ceramic humidity sensor

is a porous gold metal evaporated onto the humidity-sensing membrane. The thickness is 100–200 Å and the membrane has good moisture permeability. The electrical capacitance of the 1-μm thick sensor varies almost linearly with ambient relative humidity (*Figure 14.18*). The combination of such a delicate sensor with an insulated gate field effect transistor is normally very sensitive to drift. Electrical measurement techniques which use an alternating current signal to measure capacitance while supplying d.c. voltages of equal magnitude across the sensor to minimize the drift of mobile charges provide an accurate sensor with long time stability[15]. Details of the microchip structure and the equivalent electrical circuit are given in *Figure 14.19*. Since the d.c. resistance of the

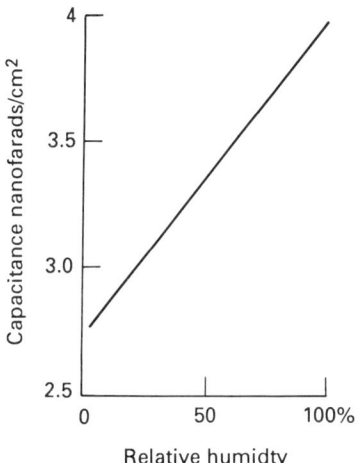

Figure 14.18 Capacitance change of cellulose acetate butyrate polymer with relative humidity

gate insulator is very high (>1000 MΩ) and the external resistance R_B is 10 MΩ, the applied d.c. voltage V_0 is equivalent to the undergate voltage on electrode G_2. However $R_B \gg 1/WC_s$ where C_s is the capacitance of the membrane. This means that the a.c. voltage on the undergate electrode G_2 depends on the impedance of the humidity-sensing membrane, which is determined by its capacitance C_s

$$\tilde{V}g = \frac{\tilde{V}o}{1 + \dfrac{C_i}{C_s}}$$

where C_i – capacitance of the gate insulator.

This electric field then controls the output of the transistor amplifier, providing an output voltage which indicates moisture[15].

14.7 Calibration

There are several laboratory techniques which are used to calibrate humidity sensors. These include mixing carefully dried air with carefully saturated air, saturating the air at a controlled temperature and raising its temperature without loss or gain of moisture, and saturating air at high pressure and then reducing the pressure to one giving the desired relative humidity[16]. However, these are cumbersome, bulky and slow and are therefore not widely used.

The widely used techniques are:

(1) *Peltier dew point determination.* This is a direct method. No calibration is needed.

(2) *Aspirated psychrometer.* High-quality wet and dry bulb thermometers are accurate and of reasonable cost. Corrections are necessary for the actual atmospheric pressure[17].

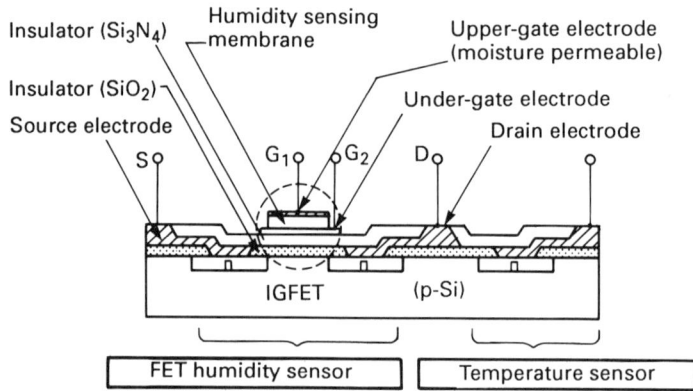

(a) Cross section of the field effect humidity sensor
(IGFET = insulated gate field effect transistor)

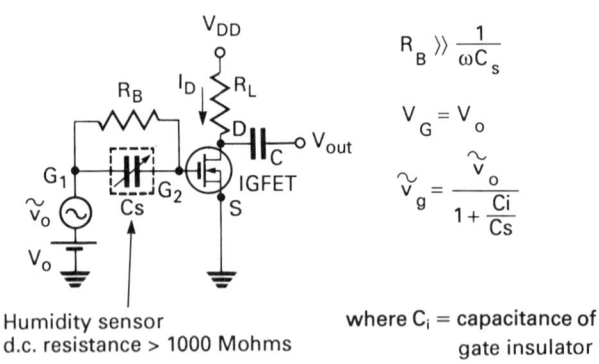

(b) Equivalent electrical circuit

Figure 14.19 Illustration of the microchip sensor

(3) *Salt solutions.* A saturated salt solution in a closed container will reach a constant relative humidity at a given temperature. The actual value of this relative humidity will vary with the salt. Some of the manufacturers of humidity sensors provide salt solutions specifically for adjusting the sensor scale. Typical salts are illustrated in *Figure 14.20*[18]. There must always be surplus solid salt present and the temperature must be stable.

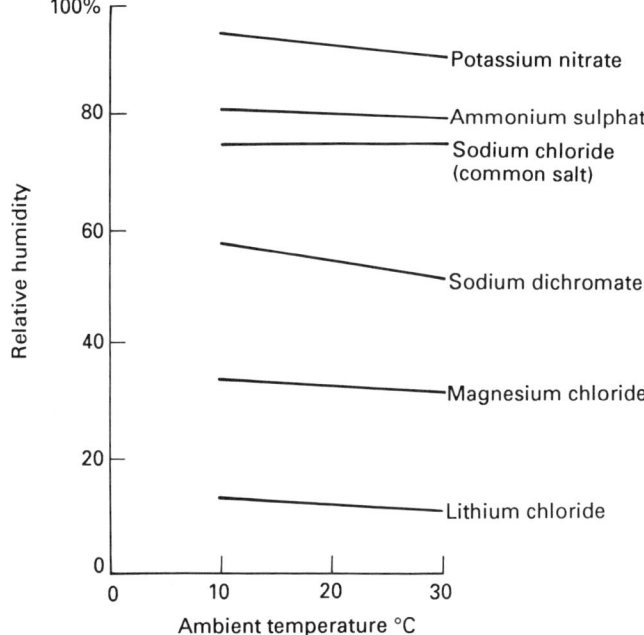

Figure 14.20 Relative humidities in equilibrium with saturated salt solutions

References

1 Leivers, M.F. and Letherman, K.M. 'The extension characteristics of some materials used for humidity sensors'. *Building Services Eng,* **45**, 205–210, 1978

2 Spencer-Gregory, H. and Rourke, E. *Hygrometry.* Crosby Lockwood, London, 1957

3 Blakemore, C. and Baker, W. 'Piezo electric moisture analysis', pp. 54–64 in *Humidity Sensors and their Calibration* (ed. W.H. McGivern). NPL, Teddington, 1986

4 Fritschen, L.J. and Gay, L.W. *Environmental Instrumentation.* Springer Verlag, New York, 1979

5 World Meteorological Organisation. *Guide to Meteorological Instrument and Observing Practice.* World Meteorological Organisation, Geneva, Switzerland, 1971

6 Lourence, F.J. and Pruitt, W.O. 'A psychrometer system for micrometeorology profile determinations'. *J Appl Meteorol,* **8**, 492–498, 1969

7 Tanner, C.B. and Suomi, V.E. 'Lithium chloride dewcel properties and use for dew point and vapour pressure gradient measurements'. *Trans Am Geophysical Union,* **37**(4), 413–420, August 1956

8 Nelson, D.E. and Amdur, E.J. 'The mode of operation of saturation temperature hygrometers based on electrical detection of salt solution phase transition', pp. 617–626 in *Humidity and Moisture: Measurement and Control in Science and Industry,* Vol. I (ed. A. Wexler). Reinhold Publishing Corporation, New York, 1965

9 Hedlin, C.P. and Trofimenkoff, F.W. 'An investigation of the accuracy and response rate of a lithium chloride heated electrical hygrometer', pp. 627–634 in *Humidity and Moisture: Measurement and Control in Science and Industry,* Vol. I (ed. A. Wexler). Reinhold Publishing Corporation, New York, 1965

10 Jumo Instruments. m.k. Juchheim Gmbh, Fulda D-6400, West Germany

11 Michell Instruments Ltd, Unit 9, Nuffield Close, Nuffield Road, Cambridge, CB4 1SS, England

12 General Eastern. *1989 Catalogue of Products.* 50 Hunt Street, Watertown, MA 02172, USA

13 Yokomizo, Y., Uno, S., Harata, M. and Hiraki, H. 'Microstructure and humidity sensitive properties of $ZnCr_2O_4$–$LiZnVO_4$ ceramic sensors'. *Sensors and Actuators*, **4**, 599–606, 1983

14 Nitta, T., Terada, Z. and Hayakawa, S. 'Humidity sensitive electrical conduction of $MgCr_2O_4$–TiO_4 porous ceramics'. *J Am Ceramic Soc*, **63**, 295–299, 1980

15 Hijikigawa, M., Sugihara, T., Tanaka, J. and Watanabe, M. 'Microchip FET humidity sensor with long term stability'. Transducers '85, Paper 9.7, pp. 221–224, DSC 10C 9020.58285, 1985

16 Middleton, W.E.K. and Spilhaus, A.F. *Meteorological Instruments*. University of Toronto Press, Canada, 1953

17 BS 4833: 1986. 'Hygrometric tables for use in the testing and operation of environmental enclosures'. British Standards Institution, London, 1986

18 Wexler, A. and Hasegawa, S. 'Relative humidity temperature relationships of some saturated salt solutions in the temperature range 0°C to 50°C'. *J Res Natl Bureau of Standards USA*, **53**, 19–26, 1957

Further reading

Abbott, N.J. 'Note on the construction of fibre hygrometers'. *Text Res J*, **24**, 59–64, 1954

Anderson, J.H. and Parks, G.A. 'The electrical conductivity of silica gel in the presence of adsorbed water'. *J Phys Chem*, **72**(10), 3662–3668, 1968

Belt, P. 'Development of polymeric humidity sensors', pp. 65–71 in *Humidity Sensors and their Calibration* (ed. W.H. McGovern). NPL, September 1986

Bennewitz, P.F. 'The Brady array: a new bulk effect humidity sensor'. *Measurements and Data*, 104–107, Jan/Feb 1972

Blakemore, C. and Baker, W. 'Piezo electric moisture analysis', pp. 54–71 in *Humidity Sensors and their Calibration* (ed. W.H. McGivern). NPL, September 1986

Buck, A.L. 'Recent developments in low frost point hygrometry', pp. 72–94 in *Humidity Sensors and their Calibration* (ed. W.H. McGivern). NPL, September 1986

Carr-Brion, K. *Moisture Sensors in Process Control*. Elsevier Applied Science Publishers, London, 1986

Cromer, C.J. 'Instrumentation for monitoring buildings in hot humid climates, 59–65. Field data acquisition for building and equipment energy use monitoring'. Natl Bur Standards Report PB 86 – 19957, 1986

Greaves, P. 'Calibration to salt systems for RH measurement', pp. 95–103 in *Humidity Sensors and their Calibration* (ed. W.H. McGivern). NPL, September 1986

Greenspan, L. 'Los frost point humidity generator'. *J Res*, **77A**(5), 671–677, 1973

Hales, J.L. 'Review of calibration methods for humidity sensors', pp. 76–82 in *Humidity Sensors and their Calibration* (ed. W.H. McGivern). NPL, September 1986

Heggemann, R.H. 'Recent developments in the self-heated lithium chloride dew point sensors'. *Analytical Instruments*, **16**, 117–120, 1978

Hurley, C.W. 'Measurement of temperature humidity and flow', pp. 43–59 in *Field data acquisition for building and equipment energy use monitoring*. Natl Bureau Standards Report PB 86 – 199957. Conf 8510218–E DE 86–008569, 1986

Lambert, L.B. 'History of humidity measurement'. *Instrument Practice*, **19**, 128–136, Feb 1965

Littler, J.W., Hasegawa, S. and Greenspan, L. 'Performance characteristic of a bulk effect humidity sensor'. NBSIR 74–477. Dept of Navy, Office of Naval Research, Virginia, 1974

McGivern, W.H. 'Humidity sensors and their calibration'. NPL Conf Proc, 17th September, 1986

Martin, S. 'The control of conditioning atmosphere by saturated salt solutions', pp. 503–507 in *Humidity and Moisture Measurement and Control in Science and Industry*, Vol. 3 (ed. A. Wexler), Reinhold Publishing Corporation, New York, 1965

Middleton, W.E.K. 'The early history of hygrometry'. *Q J R Met Soc*, **68**, 247–258, 1942

Nagamito, S., Nitta, T., Kobayashi, T. and Nakano, M. *Automatic Microwave Oven with Humidity Sensing*. Microwave Power Symposium, Ottawa, Ontario, Canada, June 1978

Nitta, T. 'Ceramic humidity sensor'. *Ind Eng Chem Prod Res Div*, **20**, 669–674, 1981

Nitta, T., Terada, Z. and Kanazawa, T. 'Ceramic humidity sensor 'Humiceram' '. *National Technical Report*, **24**(3), 422–435, 1978

Pragnell, R.F. 'Review of humidity sensors', pp. 3–11 in *Humidity Sensors and their Calibration* (ed. W.H. McGivern). NPL, September 1986

Wexler, A. (ed.). *Humidity and Moisture Control and Measurement in Science and Industry*, Vols 1 and 2. Reinhold, New York, 1965

Wexler, A. and Daniels, R.D. 'Pressure humidity apparatus'. *J Res Natl Bureau of Standards USA*, **48**(4), 269–274, 1952

Wiederhold, P.R. 'Humidity measurements. Part I. Psychrometers and per cent RH sensors'. *Instrument Technology*, **22**(6), 31–37, 1975

Wiederhold, P.R. 'Humidity measurements Part 2. Hygrometry'. *Instrument Technology*, **22**(8), 45–50, 1975

Wood, R.C. 'The infrared hygrometer – its application to difficult humidity measurement problems, pp. 492–504 in *Humidity and Moisture; Measurement and Control in Science and Industry*, Vol. 1 (ed. A. Wexler). Reinhold Publishing Corporation, New York, 1965

Definitions

There is a wide variety of ways of describing the amount of water vapour in the atmosphere. The most commonly used ones are:

(1) *Humidity mixing ratio.* This is the ratio of the mass of water vapour to the mass of dry air associated with it expressed as a percentage. It is widely used by Building Service Engineers and psychometric tables list it.

$$\text{Mixing ratio } g = \frac{\text{kg of water}}{\text{kg of dry air}}$$

(2) *Percentage saturation.* This is the ratio of the actual moisture content to the moisture content at saturation at that temperature, expressed as a percentage:

$$\text{Percentage saturation } \mu = \frac{100 \times g}{g_{ss}} \%$$

where g is the actual moisture content and g_{ss} is the moisture content of saturated air at that temperature.

(3) *Mole fraction.* The mole fraction of a given component in a mixture is equal to the number of moles of that component divided by the total number of moles of all components in the mixture.

$$\text{Mole fraction of water vapour in air } = \frac{X_w}{X_a + X_w}$$

where X_w = mole fraction of water vapour
X_a = mole fraction of air

$$\text{Moisture content } g = \frac{X_w}{X_a} \times \frac{\text{Molecular wt of water}}{\text{Molecular wt of air}}$$

$$= \frac{X_w}{X_a} \times \frac{18.01}{28.96} = 0.622 \frac{X_w}{X_a}$$

Since $X_w + X_a = 1$, this may be rewritten

$$\text{Moisture content } g = 0.622 \times \frac{X_w}{1 - X_w}$$

Mole fractions are the units used by chemists in chemical reaction analyses.

(4) *Relative humidity*. This is the ratio of the mole fraction of water vapour present to the mole fraction of water vapour at saturation for the same temperature and pressure, expressed as a percentage. Alternatively it could be considered as the ratio of water vapour pressure at present relative to that at saturation.

$$\text{Relative humidity } \phi = \frac{X_w}{X_{ws}} \times 100\%$$

where X_w = mole fraction of water vapour present
X_{ws} = mole fraction of saturated water vapour at the same temperature and pressure

Relative humidity is linked to moisture content g by the expression:

$$\text{Relative humidity } \phi = \frac{1 + (0.622/g_s)}{1 + (0.622/g)} \times 100\%$$

Relative humidity is also linked to percentage saturation μ by the expression:

$$\text{Relative humidity } \phi\% = \mu \times \frac{(0.622 + g_s)}{(0.622 + g)}\%$$

Relative humidity is the expression used by physicists. At ordinary room temperatures (20°C) their numerical values for relative humidity ϕ are very similar to, but slightly greater than, those of percentage saturation μ.
    At 50% saturation relative humidity 20°C:

$$\phi = 50 \times \frac{(0.622 + 0.01475)}{(0.622 + 0.00738)} + 50 \times 1.0117 = 50.5\%$$

(5) *Specific humidity*. The British Meteorological Office define this as the ratio of the mass of water vapour to the total mass of moist air:

$$\text{Specific humidity } q = \frac{M_w}{M_w + M_a}$$

where M_w = mass of water present
M_a = mass of air present

In terms of the humidity mixing ratio g:

$$\text{Specific humidity } q = \frac{g}{1 + g}$$

    Take care with this term. The Meteorological Office also call this term 'the moisture content'. The British Building Service engineers and their American equivalents ASHRAE also call the humidity mixing ratio g 'the moisture content'.

(6) *Vapour pressure*. The water vapour present in air behaves like any other gas and exerts a pressure which makes up part of the total pressure of the air. It is measured in millibars where one bar $= 10^5\,N/m^2$ and approximates to atmospheric pressure. Vapour pressure can be linked to the humidity mixing ratio g:

$$\text{Vapour pressure } e \;=\; \frac{g}{(0.622 + g)} \times P$$

where P = atmospheric pressure

(7) *Absolute humidity*. This is the ratio of the mass of water vapour to the total volume it occupies.

$$\text{Absolute humidity } d_v \;=\; \frac{M_w}{V}$$

where M_w = mass of water vapour
V = volume occupied

It is also known as the water vapour density.

(8) *Dew point temperature*. This is the saturation temperature for the humidity mixing ratio of the air. It is the temperature at which moisture just starts to condense out of the air stream.

(9) *Thermodynamic wet bulb temperature*. This is the temperature at which liquid water may be evaporated adiabatically into the air to bring it to saturation. The wet bulb used in psychrometry does not operate adiabatically but involves simultaneous heat and mass transfer from the bulb to the air stream. Fortunately the corrections applied to the wet bulb temperature to obtain the thermodynamic wet bulb are small.

Index